DO PENGUINS
HAVE KNEES?

ALSO BY DAVID FELDMAN

DO PENGUINS HAVE KNEES?

An Imponderables® Book

David Feldman

Illustrated by Kassie Schwan

 **Collins**

An Imprint of HarperCollins*Publishers*

For Rick Kot

A hardcover edition of this book was published in 1991 by HarperCollins Publishers.

First HarperPerennial edition published 1992.
First Perennial Currents edition published 2004.

Designed by Cassandra J. Pappas

Library of Congress Cataloging-in-Publication Data
Feldman, David.
    Do penguins have knees? / David Feldman ; illustrated by Kassie Schwan.
      p. cm.
    "An Imponderables book."
    Includes index.
    ISBN 0-06-074091-4 (pbk.)
    1. Questions and answers. 2. Questions and answers—Humor. I. Title.

AG195.F44 2004
031.02—dc22

                                             2004044786

07 08 FOLIO/RRD 20 19 18 17 16 15 14 13 12

Contents

CONTENTS   vii

CONTENTS

CONTENTS

CONTENTS

CONTENTS

Preface

You may think you've lived a happy life without knowing the answer to why you don't feel a mosquito while it's biting you. Or why the address labels on subscription magazines are usually placed upside-down. Or why the bags on oxygen masks in airplanes don't inflate.

But you're not really happy. Face it. In our everyday life, we're confronted with thousands of mysteries that we cannot solve. So we repress our anxiety (could you imagine a life haunted by the recurring dread of not knowing what function our earlobes serve?). And as those of us who've seen Dr. Joyce Brothers on television know, repression isn't good for you.

So our quest is to eradicate all of these nagging Imponderables. Luckily, we have the best possible collaborators—our readers.

Most of the Imponderables posed in this book were submitted by readers of our first four volumes. In our Frustables section, readers wrestle with the ten most frustrating Imponderables that we weren't able to solve. And in the letters section, readers suggest how our previous efforts may have been ever so slightly less than perfect.

As a token of our appreciation for your help, we offer a free copy of our next book to the first person who sends in an Imponderable or the best solution to a Frustable we use, along with an acknowledgment.

The last page of the book tells you how you can join in our campaign to stamp out Imponderability. We can't guarantee you lifelong happiness if you read *Do Penguins Have Knees?*, but we can assure you you'll know why there are peanuts in plain M&Ms.

Why Don't You Feel or See a Mosquito Bite Until After It Begins to Itch?

We would like to think that the reason we don't feel the mosquito biting us is that Mother Nature is merciful. If we were aware that the mosquito was in the process of sinking its mouth into our flesh, we might panic, especially because a simple mosquito bite takes a lot longer than we suspected.

A female mosquito doesn't believe in a casual "slam bam, thank you, ma'am." On the contrary, mosquitoes will usually rest on all six legs on human skin for at least a minute or so before starting to bite. Mosquitoes are so light and their biting technique so skillful that most humans cannot feel them, even though the insect may be resting on their skin for five minutes or more.

When the mosquito decides to finally make her move and press her lancets into a nice, juicy capillary, the insertion takes about a minute. She lubricates her mouthparts with her own saliva and proceeds to suck the blood for up to three minutes until her stomach is literally about to burst. She withdraws her lancets in a few seconds and flies off to deposit her eggs, assuring the world that the mosquito will not soon make the endangered species list.

A few sensitive souls feel a mosquito's bite immediately. But most of us are aware of itching (or in some cases, pain) only after the mosquito is long gone not because of the bite or the loss of blood but because of the saliva left behind. The mosquito's saliva acts not only as a lubricant in the biting process but as an anesthetic to the bitee. For most people, the saliva is a blessing, since it allows us to be oblivious to the fact that our blood is being sucked by a loathsome insect. Unfortunately, the saliva contains anticoagulant components that cause allergic reactions in many people. This allergic reaction, not the bite itself, is what causes the little lumps and itchy sensations that make us wonder why mosquitoes exist in this otherwise often wonderful world.

Submitted by Alesia Richards of Erie, Pennsylvania.

Why Doesn't Milk in the Refrigerator Ever Taste As Cold As the Water or Soda in the Refrigerator?

Actually, milk *does* get as cold as water or soda. If you are having a particularly boring Saturday night, you might want to stick a thermometer into the liquids to prove this.

Milk at the same temperature as water or soda just doesn't taste as cold to us because milk contains fat solids. We perceive solids as less cold than liquids. Taste experts refer to this phenomenon as "mouth feel."

If the milk/water/soda test wasn't exciting enough for you, run a test in your freezer compartment that will demonstrate the same principle. Put a pint of premium high-butterfat ice cream in the freezer along with a pint of low-fat or nonfat frozen yogurt.

DAVID FELDMAN

Consume them. We'll bet you two to one that the yogurt will taste colder than the ice cream. For the sake of research, we recently performed this experiment with due rigor, and because we wanted to go out of our way to assure the accuracy of the experiment, we conducted the test on many different flavors of ice cream and yogurt. Oh, the sacrifices we make for our readers!

Submitted by Pat O'Conner of Forest Hills, New York.

Why Are Address Labels on Subscription Magazines Usually Placed Upside-Down?

Our usually reliable sources at the United States Postal Service struck out on this Imponderable, but we were rescued by our friends at Neodata Services. Neodata, the largest fulfillment house in the United States, which we profiled in *Why Do Dogs Have Wet Noses?*, is the company that processes all those subscription forms you send to Boulder, Colorado.

By luck, we rang up Neodata's Biff Bilstein when he was in a meeting with sales executives Mark Earley and Rob Farson. The three share over seventy-five years of experience in the magazine business. "So," we implored, "why are address labels placed upside-down?"

They conferred and answered as one. Even though the folks at the USPS don't seem to know it, the labels are placed upside down to accommodate the postal carrier. All magazines are bound on the left-hand side. Our hypothetical postal carrier, being right-handed, naturally picks up a magazine by the spine with his or her right hand to read the address label—the magazine is thus automatically turned upside down. But the label is now "right side up" and easily read by the postal carrier. Nifty, huh?

Submitted by Geoff Grant of Barrie, Ontario. Thanks also to Beth Jones of West Des Moines, Iowa.

Why Are There Dents on the Top of Cowboy Hats?

Of course, not *all* cowboy hats have dents. How about country and western star George Strait's? Or *Bonanza*'s Dan ("Hoss") Blocker's?

Yet the vast majority of cowboy hats do have dents, and no one we spoke to could give us any other explanation than that dents are there "for style." Ralph Beatty, director of the Western/English Retailers of America, theorizes that early cowboy hats may have acquired dents by wear, and later were intentionally added.

As one, better-to-be-kept-anonymous, western hat marketer put it, "Let's face it. Without the dent, you would look like a dork."

We wonder if he would have said that to Dan Blocker's face.

Submitted by Lisa R. Bell of Atlanta, Georgia.

DAVID FELDMAN

Why Do Grocery Coupons State That They Have a "Cash Value of 1/100 of 1¢"?

We receive a lot of questions not only about everyday life but about the questions we get asked. The most frequently asked question about questions: What is your most frequently asked question?

Imponderables run in cycles. After our first book, *Imponderables*, was published, "Why are buttons on men's shirts and jackets arranged differently from those on women's shirts?" was the most popular question. Then it was "Why is yawning contagious?" Then "Why can't we tickle ourselves?" The all-time champion, though, is "Why do we park on driveways and drive on parkways?" We don't know why people care passionately about this subject, but this is *the* Imponderable that just will not go away. We've discussed the answer in two books, and it is still our second most frequently asked Imponderable.

But the clear champion now is the Imponderable at hand. More than thirty readers have asked this Imponderable in the last two years, and the irony is that the question was one of the original Imponderables we hoped to answer in our first book.

We have spoken to scores of officials in the coupon processing, direct marketing, grocery, and marketing science fields, but nobody could pinpoint the exact reason for the custom or for the particular price of 1/100 of one cent. To make matters worse, we then received a follow-up letter from Kathy Pierce, one of the legions who asked us this question:

> I was reading *The Straight Dope* by Cecil Adams (I'm sorry, I would guess that old Cec' is your arch enemy, but I have to have something besides my Italian textbook to read in between *Imponderables* books). Lo and behold, there was *my* question, right there on page 329.

Kathy proceeded to cite Adams's answer, and we noted that "old Cec'," whom we like to think of as a colleague and pal (at least when there are other people around to hear us think), got just about as far as we had in our research. As usual, he was depressingly accurate.

He noted that some states have laws equating coupons with trading stamps (e.g., S&H Green Stamps). Since consumers pay for the "free" stamps in the form of higher prices for groceries, the jurisdictions forced stamp issuers to redeem the stamps for a cash value. In order to comply with these state laws, which were actually designed to curb abuses among trading stamp issuers, coupon issuers assigned a cash value that nobody in his or her right mind would bother to collect.

We make it a policy to try not to repeat questions or answers we've seen discussed elsewhere (after all, if we know the answer to a question, it's not an Imponderable anymore), but since this is such a popular question, and since we have STARTLING NEW INFORMATION, we are pleased to disclose the true story behind the coupon cash value.

Although some other states treat coupons as scrip, the real driving force behind the practice is the state of Kansas. Ed Dunn, a spokesperson for NCH Promotional Services, told us what prompted these state laws. During the Depression, many stamp issuers would claim that their books of stamps were worth much more than they really were. They would then sell merchandise through catalogs at greatly inflated prices.

This caused problems. Because both the "cash value" and redemption prices (in stamps) were greatly inflated, honest stamp issuers were at a competitive disadvantage, because their own books of stamps didn't seem to be worth much in buying power compared to those of others.

Several states tried to eliminate these injustices by making all books of stamps, *and anything of value that might be used to reduce the price of a product,* have a common value. Obviously, coupons fell into this category.

But Kansas enacted by far the most stringent law. Kansas law overrides the terms and conditions of the coupon for residents of the state and, more important for our purposes, says that *if no cash value is stated on the coupon, the consumer may*

DAVID FELDMAN

cash in the coupon at face value. Obviously, if consumers could take the fifty cents they "save" on their laundry detergent and redeem it for cash, they would. Manufacturers had two choices: make separate coupons for Kansas, or state a cash value on every coupon.

Do folks really try to redeem coupons for the lofty sum of 1/100 of a cent apiece? Not very often (of course, that's why the cash value is set so low). Most companies will probably pay the tariff, but the consumer is stuck with postage costs, which far exceed the refund.

Submitted by Jeff Burger of Phoenix, Arizona. Thanks also to Lisa Lindeberg of Van Nuys, California; Anand S. Raman of Oak Ridge, Tennessee; Larry Doyle of Grand Ledge, Michigan; Jonathan Sabin of Bradenton, Florida; Brian J. Sullivan of Chicago, Illinois; Randy S. Poppert of Willow, Arkansas; Maria Scott of Cincinnati, Ohio; Joe Crandell of Annandale, Virginia; David Grettler of Newark, Delaware; and many others. Special thanks to Kathy Pierce of Boston, Massachusetts.

How and Why Did 7UP Get Its Name?

7UP (a.k.a. Seven-Up) was the brainchild of an ex-advertising and merchandising executive, C. L. Grigg. In 1920, Grigg formed the Howdy Company in St. Louis, Missouri, and found success with his first product, Howdy Orange drink.

Intent upon expanding his empire, Grigg spent several years testing eleven different formulas of lemon-flavored soft drinks. In 1929, he introduced Seven-Up, then a caramel-colored beverage.

So where did the "7" and "UP" come from? Despite its identification as a lemon-lime drink, 7UP is actually a blend of seven natural flavors. According to Jim Ball, vice-president of corporate communications for Dr Pepper/Seven-Up Companies,

Inc., all of the early advertisements for the new drink described a product that was uplifting and featured a logo with a winged 7. Long before any caffeine scares, 7UP was promoted as a tonic for our physical and emotional ills:

> Seven-Up energizes—*sets you up*—dispels brain cobwebs and muscular fatigue.

> Seven-Up is as pure as mountain snows . . .

> Fills the mouth [true, but then so does cough syrup]—thrills the taste buds—cools the blood—energizes the muscles—soothes the nerves—and makes your body alive—glad—happy.

7UP's advertising has improved and changed markedly over the years, but its name has proved to be durably effective, even if customers don't have the slightest idea what "Seven-Up" means. Grigg could have chosen much worse. Contemplate sophisticated adults sidling up to a bar and ordering a bourbon and Howdy Lemon-Lime drink.

> *Submitted by Richard Showstack of Newport Beach, California. Thanks also to Roya Naini of Olympia, Washington; Brian and Ingrid Aboff of Beavercreek, Ohio; and Jason M. Holzapfel of Gladstone, Missouri.*

Why Do the Back Wheels of Bicycles Click When You Are Coasting or Back Pedaling?

Has there ever been a child with a bicycle who has not pondered this Imponderable? We got the scoop from Dennis Patterson, director of import purchasing of the Murray Ohio Manufacturing Co.:

> The rear sprocket cluster utilizes a ratchet mechanism that engages during forward pedaling, but allows the rear wheel to rotate independently of the sprocket mechanism. When one

DAVID FELDMAN

ceases to pedal, the wheel overrides the ratchet and the clicking noise is the ratchets falling off the engagement ramp of the hub.

The ramp is designed to lock engagement if pedaled forward. The ratchet mechanism rides up the reverse slope and falls off the top of the ramp when you are coasting or back pedaling.

Submitted by Harvey Kleinman and Merrill Perlman of New York, New York.

Why Do Male Birds Tend to Be More Colorful Than Females? Is There Any Evolutionary Advantage?

"Sexual dimorphism" is the scientific term used to describe different appearances of male and female members of the same species. Charles Darwin wrestled with this topic in his theory of sexual selection. Darwin argued that some physical attributes of birds evolved solely to act as attractants to the opposite sex. How can you explain the train of the peacock except to say it is the avian equivalent of a Chippendale's dancer's outfit? As Kathleen Etchpare, associate editor of the magazine *Bird Talk*, put it: "As far as an evolutionary advantage goes, the mere number of birds in the world today speaks for itself."

Sure, but there are plenty of cockroaches around today, too, and they have managed to perpetuate themselves quite nicely without benefit of colorful males. Many ornithologists believe that the main purpose of sexual dimorphism is to send a visual message to predators. When females are nesting, they are ill-equipped to fend off the attacks of an enemy. Michele Ball, of the National Audubon Society, says that "It behooves the female to be dully colored so that when she sits in the nest she is less conspicuous to predators."

Conversely, the bright plumage of many male birds illustrates the principle of "the best offense is a good defense." Male birds, without the responsibility of nesting, and generally larger in size than their female counterparts, are better suited to stave off predators. The purpose of their bright coloring might be to warn predators that they will not be easy prey; most ornithologists believe that birds are intelligent enough to register the dimorphic patterns of other birds.

And most animals are as lazy as humans. Given a choice, predators will always choose the easy kill. If a predator can't find a dully colored female and fears the brightly colored male, perhaps the predator will pick on another species.

Submitted by Karen Riddick of Dresden, Tennessee.

DAVID FELDMAN

What Does the USPS Do with Mail It Can't Deliver or Return Because of Lack of a Return Address?

If a piece of mail is improperly addressed and does not contain a return address, it is sent to a dead letter office. Dead letter offices are located in New York, Philadelphia, Atlanta, San Francisco, and St. Paul. There a USPS employee will open the envelope. If no clues to the address of the sender or receiver are found inside, and the enclosures are deemed to have "no significant value," the letter is destroyed immediately.

Frank Brennan, of the USPS media relations division, explains that if the enclosures are deemed to be of some value, the parties involved will have a temporary reprieve:

> This allows time for inquiries and claims to be filed. After 90 days, all items that have not been claimed are auctioned off to the public. Cash or items of monetary value that are found in the mail are placed into a general fund. If it is not claimed after one year,

it is rolled over into a USPS account to be used as the USPS deems necessary.

And of course we can all count on the USPS making the best possible use of any windfalls that come their way.

Submitted by Kathryn Rehrig of Arlington, Texas.

Why Are Baseball Dugouts Built So That They Are Half Below Ground?

If dugouts were built any higher, notes baseball stadium manufacturer Dale K. Elrod, the sight lines in back of the dugout would be blocked. Baseball parks would either have to eliminate choice seats behind the dugout or sell tickets with an obstructed view at a reduced price.

If dugouts were built lower, either the players would not be able to see the game without periscopes or they wouldn't have room to stretch out between innings.

Submitted by Alan Scothon of Dayton, Ohio.

Why Do Trains with More Than One Locomotive Often Have One (or More) of the Locomotives Turned Backwards?

Diesel locomotives work equally well traveling in either direction. Robert L. Krick, deputy associate administrator for technology development at the Federal Railroad Administration, wrote *Imponderables* that

> Locomotives are turned on large turntables, or on "wye" or "loop" tracks. Railroads avoid unnecessary turning of locomotives

DAVID FELDMAN

because the procedure takes time. The locomotives being turned and the employees turning them could be employed for more constructive purposes.

When locomotives are assembled for a train, if one already faces forward it is selected for the lead position. The others will work equally well headed in either direction; they are usually coupled together without regard for their orientation.

If a group of locomotives is assembled for more than one trip, the cars will often be arranged with the rear locomotive of the group facing the rear. That group of locomotives can then be used on another train going in either direction without any turning or switching.

Using this method, a train can be returned to its original destination on the same track without any turning. Bob Stewart, library assistant at the Association of American Railroads, explains how:

When a train reaches the end of its run and is to return in the direction from which it came, the engineer moves to the cab at the other end. The locomotive can be coupled and switched to a parallel track, run back towards what was the rear of the train and switched back to the original track.

Submitted by Randy W. Gibson of Arlington, Virginia.

Why Is There Steam Coming Up from the Streets of New York?

Historically, of course, the rising steam served the most important purpose of providing menacing atmosphere in *Taxi Driver* and other movies set in New York City. But we still see steam rising out of manhole covers in Manhattan all the time. What causes it?

The biggest source of the steam is New York's utility, Consolidated Edison, which still generates enough steam to service over 2,000 customers in Manhattan. Steam heat is used only in tall buildings and manufacturing plants; the equipment necessary to generate steam power is too large and inefficient for small businesses or modest residential dwellings.

When a small leak occurs in a steam pipe, the vapor must go somewhere. Heat rises and looks for a place to escape: Manhole covers are the most likely egress point for steam. Martin Gitten, Con Edison's assistant director for public information, told *Imponderables* that when a big leak occurs, Con Ed must put tall

DAVID FELDMAN

cones over the manhole covers so that the steam is vented above the level of vehicles. Otherwise, unsuspecting drivers would feel as if they were driving through a large cumulus cloud.

Another source for steam rising out of the streets of New York is excess moisture condensing underground. The excess moisture may emanate from small leaks in city water mains, run-off from heavy rainfalls, or least pleasant to contemplate, sewer backups.

Why do these liquids rise up as steam? Because they come in contact with the scalding hot steam equipment below ground.

Submitted by Chris McCann of New York, New York.

HOW Did They Keep Beer Cold in the Saloons of the Old West?

Just about any way they could. In the nineteenth century, guzzlers didn't drink beer as cold as they do now (the English often imbibed pints of ale warm, for goodness' sake, and still do—as do the Chinese) but even grizzled cowboys preferred their brew cool.

In colder areas of the West, saloons used to gather ice from frozen lakes in the winter. John T. McCabe, technical director of the Master Brewers Association of the Americas, says that the harvest was stored in ice houses, "where the blocks of ice were insulated with sawdust. This method would keep ice for months."

Even where it wasn't cold enough for ice to form, many saloons in the Old West had access to cool mountain streams. Historical consultant William L. Lang wrote *Imponderables* that saloon workers would fill a cistern with this water to store and cool barrels of beer.

And if no cold mountain stream water was available? Phil Katz, of the Beer Institute, says that up until about 1880, many saloons built a root cellar to house beer. Usually built into the side of a hill, root cellars could keep beer below 50 degrees Fahrenheit.

And what if you wanted cold beer at home? According to Lang, "Beer was served in buckets or small pails, and often kids delivered the beer home from the saloons." Consumers in the mid-nineteenth century thought no more of bringing home "take-out beer" than we would think of ordering take-out Chinese food.

Beer expert W. Ray Hyde explains that we needn't feel sorry for the deprivation of Old Westerners before the days of refrigeration. In fact, those might have been the "good old days" of American beer:

> Beer in the Old West wasn't cold in the modern sense of the word—but it was refreshingly cool. Evaporation kept it that way. Beer in those days was packaged in wooden barrels, and the liquid would seep through the porous wood to the outside of the barrel, where it would evaporate. And basic physics explains the cooling effect of evaporation.
>
> Also, it should be noted that beer then was not artificially carbonated. The slight natural carbonation required only that it be cool to be refreshing and tasty. Modern beer, with its artificial carbonation, needs to be very cold to hide the sharp taste of the excess carbon dioxide.

Submitted by Dr. Robert Eufemia of Washington, D.C.

DAVID FELDMAN

What Is the Official Name of the Moon?

Along with our correspondent, we've never known what to call our planet's satellite. Moon? The moon? moon? the moon? Dorothy?

We know that other planets have moons. Do they all have names? How do astronomers distinguish one moon from another?

Whenever we have a problem with matters astronomical, we beg our friends at two terrific magazines—*Astronomy* and *Sky & Telescope*—for help. As usual, they took pity on us.

Astronomy's Robert Burnham, like most senior editors, is picky about word usage:

> The proper name of our sole natural satellite is "the Moon" and therefore . . . it should be capitalized. The 60-odd natural satellites of the other planets, however, are called "moons" (in lower case) because each has been given a proper name, such as Deimos, Amalthea, Hyperion, Miranda, Larissa, or Charon.

Likewise, the proper name for our star is "the Sun" and that for our planet is "Earth" or "the Earth." It's OK, however, to use "earth" in the lower case whenever you use it as a synonym for "dirt" or "ground."

Alan MacRobert, of *Sky & Telescope*, adds that Luna, the Moon's Latin name, is sometimes used in poetry and science fiction, but has never caught on among scientists or the lay public: "Names are used to distinguish things from each other. Since we have only one moon, there's nothing it needs to be distinguished from."

Submitted by A. P. Bahlkow of Sudbury, Massachusetts.

DAVID FELDMAN

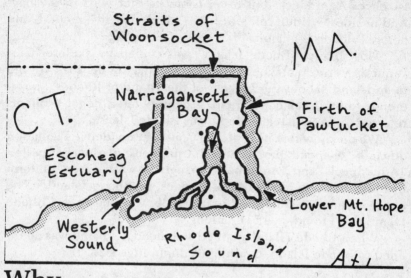

Straits of Woonsocket

MA.

CT.

Narragansett Bay

Firth of Pawtucket

Escoheag Estuary

Westerly Sound

Rhode Island Sound

Lower Mt. Hope Bay

Atl

Why Is Rhode Island Called an Island When It Obviously Isn't an Island?

Let's get the island problem licked first. No, technically, the whole state isn't an island, but historians are confident that originally "Rhode Island" referred not to the whole territory but to what we now call Aquidneck Island, where Newport is located. Christine Lamar, an archivist for the Rhode Island State Archives, endorses this view.

Why "Rhode"? Lame theories abound. One is that the state was named after a person named Rhodes (although any meaningful details about this person are obscure). Another supposition is that "Rhode Island" was an Anglicization of "Roode Eyelandt," Dutch for "red island." The Dutch explorer Adriaen Block noted the appearance of a reddish island in the area, and maps of the mid-seventeenth century often refer to the area as "Roode Eyelandt."

But all evidence points to the fact that Block was referring not to the landlocked mass of Rhode Island, nor even to the island of Aquidneck, but to an island farther west in the bay. And besides, written references to "Rhode Island" abound long before "Roode Eyelandt."

Most likely, "Rhode Island" was coined by explorer Giovanni da Verrazano, who referred in his diary of his 1524 voyage to an island "about the bigness of the Island of Rhodes," a reference to its Greek counterpart. A century later, Roger Williams referred to "Aqueneck, called by us Rhode Island . . ."

We do know that in 1644, the Court of Providence Plantation officially changed the name of Aquidneck (variously spelled "Aquedneck" and "Aquetheck"—spelling was far from uniform in those days) to "The Isle of Rhodes, or Rhode Island." The entire colony, originally settled in 1636, was known as "Rhode Island and Providence Plantations."

When Rhode Island attained statehood, its name was shortened to Rhode Island, befitting its diminutive size.

Submitted by Tony Alessandrini of Brooklyn, New York. Thanks also to Troy Diggs of Jonesboro, Arkansas.

Why Do Blacktop Roads Get Lighter in Color As They Age?

Our correspondent ponders:

When fresh blacktop roads are laid, they are pure black. Why is it that after a few years, they turn gray? You can notice this when they patch potholes. The filler material is a dark contrast to the surrounding road. Even last year's patched potholes are grayer than the new blacktop patches.

You would think dirt and "worn rubber dust" would make the road blacker, not lighter.

There is only one flaw in your question, Bill. Blacktop isn't pure black, as Amy Steiner, program director of the American

Association of State Highway and Transportation Officials, explains:

> The primary ingredients in "blacktop" are asphalt and stones. Asphalt coats the stones and gives the pavement its black color. As traffic passes over the pavement, the asphalt coating on the surface stones wears off. Since stones are generally lighter in color than asphalt, the road becomes lighter in color.

The other main reason that blacktop lightens in color is oxidation. As the road surface is always exposed to the ambient air, it naturally becomes lighter.

As for what happens to the black tire tread that comes off vehicles, may I suggest you read a stimulating, brilliantly written dissertation on the subject in a wonderful book, *Why Do Clocks Run Clockwise? and Other Imponderables*. The name of the author escapes us at the moment, but we're sure your local bookstore employee will happily lead you to the HarperPerennial book, which, we recollect, is very reasonably priced.

Submitted by Bill Jelen of Akron, Ohio.

Why Must We Push Both the "Record" and "Play" Switches to Record on an Audio Tape Recorder, and Only the "Record" on the VCR?

All of our electronics sources agreed that consumers prefer "one-touch recording" for both audio and video recorders. All agreed that there is no difference in the performance of decks with one-touch or two-touch controls. So why do we have to go the extra step on the audio recorder? Audio recorders predate video recorders, and the history of the audio tape deck gives us our answer. Thomas Mock, director of engineering for the Electronics Industries Association, explains:

> In most earlier audio recorders, the switches were mechanically coupled to the tape drive mechanism. The RECORD button

was designed so that it was not possible to accidentally go into the record mode while playing a tape. In order to record, the RECORD button had to be depressed first, allowing it to sense if the cassette ERASE tab was present. Then it would permit closure of the play button.

Modern recorders, audio as well as VCRs, use servo controls to engage the tape mechanism and sensors to detect if erasure/recording is allowed. With these devices, "one button" recording is possible from the control panel or via an infrared remote or by a preset timer.

William J. Goffi, of the Maxell Corporation, told *Imponderables* that many audio recorder manufacturers have seen the light and are incorporating user-friendly one-touch recording.

Submitted by Richard Stans of Baltimore, Maryland.

How Do Bus Drivers Get into a Bus When the Door Handle Is Inside the Bus?

It all depends upon the bus. According to Robin Diamond, communications manager of the American Bus Association, many newer buses have a key lock that will open the door automatically when a key is turned. Mercedes-Benz buses, according to their press information specialist, John Chuhran, have a hydraulic door release that can be activated "by a key located in an inconspicuous place." Most often, the "inconspicuous place" is the front of the bus rather than the door itself.

Instead of a key-activated mechanism, some buses have a handle or air-compression button located in the front of the bus. Others have a toggle switch next to the door that opens it. Along with these high-tech solutions, we heard about some other strategies for bus drivers who may have locked themselves out. Karen E. Finkel, executive director of the National School Transportation Association, was kind enough to supply them:

DAVID FELDMAN

1. Enter through the rear emergency door, which does have a handle.
2. Push the door partially closed, but not enough for the door mechanism to catch, so that the door can be pulled open.
3. Use your hands to pry open the door.

Why do we think that method #3 is used altogether more often than it is supposed to be?

Submitted by Harry C. Wiersdorfer of Hamburg, New York.
Thanks also to Natasha Rogers of Webster, New York.

Why Is the Lowest-Ranked Admiral Called a *Rear* Admiral?

If you think that we are going to joke about the fact that a rear admiral is the lowest-ranked admiral because he tends to sit on his duff all day, you severely underestimate us. Puns are the refuge of the witless.

Dr. Regis A. Courtemanche, professor of history at the C. W. Post campus of Long Island University, wrote to *Imponderables* that the term originally referred to the admirals who commanded English naval fleets in the seventeenth-century Dutch Wars. The fleets were divided into three segments: the vanguard (the ships in front), the center, and the rear. "So," Courtemanche concludes, "the term lies in the fact that the *lowest* ranking admiral controlled the *rear* of the fleet at sea."

Submitted by Peter J. Scott of Glendale, California.

Why Was April 15 Chosen as the Due Date for Taxes?

It wasn't ever thus. In fact, the original filing date for federal taxes, as prescribed in the Revenue Act of 1913, was March 1. A mere five years later, the deadline moved back; until the Internal Revenue Code of 1954 was approved on August 16, 1954, midnight vigils were conducted on *March* 15. Taxpayers who paid on a fiscal year were also given a month's extension in 1954, so that they now filed on the fifteenth day of the fourth month, instead of the third month, after their fiscal year was over. In fact, all federal returns, with the exception of estates and trusts, are now due on April 15, or three and one-half months after the end of the fiscal year.

Were these dates plucked out of thin air? Not really. The IRS wants to process returns as early in the year as possible. In the 1910s, when most tax returns were one page long, it was assumed that after a wage earner totaled his or her income, the return could be filled out in a matter of minutes. Why wait until after March 1? As anyone who now is unfortunate enough to make a so-called living knows, the IRS form isn't quite as simple as it used to be. The 1040 is no easier to decipher than the Dead Sea Scrolls.

Kevin Knopf, of the Department of the Treasury, was kind enough to send us transcripts of the hearings before the House Ways and Means Committee in 1953 pertaining to the revision of the Internal Revenue Service Code. Now we don't necessarily expect the contents of all hearings in the legislature to match the Lincoln-Douglas debates in eloquence and passion, but we were a little surprised to hear the original impetus for the legislation cited by a sponsor of the 1954 IRS revision, the Honorable Charles E. Bennett of Florida, who argued for changing the due date from March 15 to April 15:

> The proposal to change the final return date from March 15 to April 15 was first called to my attention by the Florida Hotel Association. They advised that many taxpayers must cut their winter vacations short to return to their homes and to prepare their tax returns for filing before March 15. They pointed out that changing the deadline to April 15 would help their tourist trade as well as that of other winter tourist areas in the United States such as California, Arizona, Maine, and Vermont.

This is why we changed the tax code? Probably not. A succession of witnesses before the House Ways and Means Committee—everybody from the Georgia Chamber of Commerce to the American Federation of Labor to the American Cotton Manufacturers Institute—argued for moving back the date of tax filing. In descending order of importance, here were their arguments.

1. Taxpayers need the extra time to compile their records and fill out the tax forms.
2. The IRS needs more time to process returns efficiently. If the date were moved back to April 15, the IRS could rely more on permanent employees rather than hiring temporary help during

the crunch. Perhaps so many taxpayers wouldn't file at the dead-line date if they had an extra month.

    3. An extension would also ease the task of accountants and other tax preparers.

    4. It would make it easier for people who have to estimate their tax payments for the next year to make an accurate assessment.

    5. It would allow businesses who have audits at the end of the year time to concentrate on their IRS commitments.

The 1954 bill passed without much opposition. The April 15 date has proved to be workable, but it is no panacea. Any fantasy that most taxpayers wouldn't procrastinate until the last minute was quickly dispelled.

This drives the IRS nuts, because most taxpayers receive refunds. The basis of the free-market economic system is supposed to be that people will act rationally in their economic self-interest. If this were true, taxpayers with refunds would file in January in order to get their money as fast as possible, since the IRS does not pay interest on money owed to the taxpayer.

The IRS would love to find a way to even out its workload from January through April. In reality, most returns are filed either in late January and early February or right before the April 15 deadline. A 1977 internal study by the IRS, investigating changing the filing dates, said that "These peaks are so pronounced that Service Centers frequently have to furlough some temporary employees between the two workload peaks."

Before the code changed in 1954, the IRS experienced the same bimodal pattern—the only difference was that the second influx occurred in mid-March instead of mid-April. If the due date was extended a month, the second peak would probably occur in mid-May.

The IRS has contemplated staggering the due dates for different taxpayers, but the potential problems are huge (e.g., employers would have to customize W-2s for employees; single filers who get married might end up with extra-short or extra-long tax years when they decided to file a joint return; if a change

                    DAVID FELDMAN

in the tax rate occurs, when does it take effect?; would states and cities conform to a staggered schedule?) and probably not worth the effort. The same study contemplated extending the filing date (while offering financial incentives for filing early) but also concluded that the potential traps outweigh the benefits.

The IRS knows that many taxpayers deliberately overwithhold as a way to enforce savings, even though they will not collect any interest while the government holds their money. These overwithholders, flouting the advice of any sensible accountant, are most unlikely to be tempted to file early because of a possible $10 bonus from the IRS.

Now that the IRS grants an automatic two-month extension on filing to anyone who asks for it, even tax preparers are generally against changing the April 15 deadline. Henry W. Bloch, the president of H&R Block, has penetrated the very soul of his customer, and in 1976 offered this appraisal in the *Kansas City Times:*

> We get people in our office at 10 or 11 the night of April 15 and then they run down to the post office. If you extended that April 15 deadline to June 30, in my opinion, all they're going to do is wait until June 30 instead of April 15. . . . The reason for that is simply the old American habit of putting things off.

Submitted by Richard Miranda of Renton, Washington. Thanks. also to Edward Hirschfield of Portage, Michigan.

What's the Difference Between a Lake and a Pond?

"This is an Imponderable?" we hear you muttering beneath your breath as you read the question. "A lake is a big pond."

Sure, you're right. But have you considered exactly what is the dividing line in size between a lake and a pond? And what separates a lake from a sea or a pool? Do you think you know the answer?

Well, if you do, why don't you go into the field of geography or topography or geology? Because the professionals in these fields sure don't have any standard definitions for any of these bodies of water.

As stated in the *National Mapping Division's Topographic Instructions'* "Glossary of Names for Topographic Forms," a lake is "Any standing body of inland water generally of considerable size." The same publication classifies a pond as "a small fresh-water lake." But other government sources indicate that salt-water pools may be called lakes.

And absolutely no one is willing to say what the dividing line in size is between the lake and the pond. In fact, the only absolutely clear-cut distinction between the two is that a lake is always a natural formation; if it is manmade, the body is classified as a pond. Ponds are often created by farmers to provide water for livestock. Some ponds are created to provide feeding and nesting grounds for waterfowl. Hatcheries create stocked ponds to breed fish.

Many communities try to inflate the importance of their small reservoirs by calling them lakes rather than ponds. No one is about to stop them.

> *Submitted by Jeffrey Chavez of Torrance, California. Thanks also to Ray Kerr of Baldwin, Missouri, and Eugene Bender of Mary, Missouri.*

What's the Difference Between an Ocean and a Sea?

The same folks who are having trouble distinguishing between lakes and ponds are struggling with this one, too. Once again, there is general agreement that an ocean is larger than a sea.

The standard definition of an ocean, as stated in the United States Geological Survey's Geographic Names Information Service, is "The great body of salt water that occupies two-thirds of

DAVID FELDMAN

the surface of the earth, or one of its major subdivisions." Notice the weasel words at the end. Is the Red Sea a "major subdivision" of the Indian Ocean? If so, why isn't it the Red Ocean? Or simply referred to as the Indian Ocean?

Most, but by no means all seas are almost totally landlocked and connected to an ocean or a larger sea, but no definition we encountered stated this as a requirement for the classification. Geographical and geological authorities can't even agree on whether a sea must always be saline: the United States Geological Survey's Topographical Instructions say yes; but in their book *Water and Water Use Terminology*, Professors J. O. Veatch and C. R. Humphrys indicate that "sea" is sometimes used interchangeably with "ocean":

> In one place a large body of salt water may be called *lake*, in another a *sea*. The Great Lakes, Lake Superior and others, are fresh water but by legal definition are *seas*.

The nasty truth is that you can get away with calling most places whatever names you want. We often get asked what the difference is between a "street" and an "avenue" or a "boulevard." At one time, there were distinctions among these classifications: A street was a paved path. "Street" was a useful term because it distinguished a street from a road, which was often unpaved. An avenue was, in England, originally a roadway leading from the main road to an estate, and the avenue was always lined with trees. Boulevards were also tree-lined but were much wider thoroughfares than avenues.

Most of these distinctions have been lost in practice over the years. Developers of housing projects have found that using "street" to describe the roadways in their communities makes them sound drab and plebeian. By using "lane," which originally referred to a narrow, usually rural road, they can conjure up Mayberry rather than urban sprawl. By using "boulevard," a potential buyer visualizes Paris rather than Peoria.

For whatever reason, North Americans seem to like lakes more than seas. We are surrounded by oceans to the west and east. By standard definitions, we could certainly refer to Lake

Ontario, which is connected, via the St. Lawrence, to the Atlantic, as the Ontario Sea. But we don't. And no one, other than *Imponderables* readers, evidently, is losing any sleep over it.

Submitted by Don and Marian Boxer of Toronto, Ontario.
Thanks also to June Puchy of Lyndhurst, Ohio.

Why Does the United States Mint Use a Private Firm— UPS—to Ship Its Coin Sets?

Why would anyone, even a governmental agency, want to use the boringly reliable United Parcel Service when it could experience the excitement and sense of danger in using the United States Postal Service to ship its coin sets? By using the USPS, every order's fate could be a potentially unsolved mystery.

Of course, every governmental agency has its own budget to worry about. If a government office feels it can save money or save time by using private industry, it is under no obligation to throw its business to a government agency.

The U.S. Mint actually does use the USPS to ship some coin orders. Francis B. Frere, assistant director of the Mint for sales operations, explained the Department of the Treasury's policy:

> In making the determination as to which service to use, we look at the product we are shipping and the cost involved, taking into consideration value, weight, and distance.
>
> Cost is a concern to us. There are substantial savings to be realized in shipping coins by UPS. On an annual basis, we achieve savings in excess of $1 million by selectively shipping our products by UPS rather than first class mail through the U.S. Postal Service. UPS insures all packages against loss or damage.
>
> The Mint's coin programs are self-supporting. It is our responsibility to manage the coin programs in the most effective and economical manner possible . . .

Submitted by Ray W. Cummings of St. Louis, Missouri.

DAVID FELDMAN

Technical Merit 5.6
Artistic Impression 5.7
Dizziness Achieved

How Do Figure Skaters Keep from Getting Dizzy While Spinning? Is It Possible to Eye a Fixed Point While Spinning So Fast?

Imponderables readers aren't the only ones interested in this question. So are astronauts, who suffer from motion sickness in space. We consulted Carole Shulman, executive director of the Professional Skaters Guild of America, who explained:

> Tests were conducted by NASA several years ago to determine the answer to this very question. Research proved that with a trained skater, the pupils of the eyes do not gyrate back and forth during a spin as they do with an untrained skater. The rapid movement of the eyes catching objects within view is what actually causes dizziness.
>
> The eyes of a trained skater do not focus on a fixed point during a spin but rather they remain in a stabilized position focus-

ing on space between the skater and the next closest object. This gaze is much like that of a daydream.

So how are skaters taught to avoid focusing on objects or people in an arena? Claire O'Neill Dillie, skating coach and motivational consultant, teaches students to see a "blurred constant," an imaginary line running around the rink. The imaginary line may be in the seats or along the barrier of the rink (during layback spins, the imaginary line might be on the ceiling). The crucial consideration is that the skater feels centered. Even when the hands and legs are flailing about, the skater should feel as if his or her shoulders, hips, and head are aligned.

Untrained skaters often feel dizziest not in the middle of the spin but when stopping (the same phenomenon experienced when a tortuous amusement park ride stops and we walk off to less than solid footing). Dillie teaches her students to avoid vertigo by turning their heads in the opposite direction of the spin when stopping.

What surprised us about the answers to this Imponderable is that the strategies used to avoid dizziness are diametrically opposed to those used by ballet dancers, who use a technique called "spotting." Dancers consciously pick out a location or object to focus upon; during each revolution, they center themselves by spotting that object or location. When spotting, dancers turn their head at the very last moment, trailing the movement of the body, whereas skaters keep their head aligned with the rest of their body.

Why won't spotting work for skaters? For the answer, we consulted Ronnie Robertson, an Olympic medalist who has attained a rare distinction: Nobody has ever spun faster on ice than him.

How fast? At his peak, Robertson's spins were as fast as six revolutions per second. He explained to us that spotting simply can't work for skaters because they are spinning too fast to focus visually on anything. At best, skaters are capable of seeing only the "blurred constant" to which Claire O'Neill Dillie was referring, which is as much a mental as a visual feat.

DAVID FELDMAN

Robertson, trained by Gustav Lussi, considered to be the greatest spin coach of all time, was taught to spin with his eyes closed. And so he did. Robertson feels that spinning without vertigo is an act of mental suppression, blocking out the visual cues and rapid movement that can convince your body to feel dizzy.

Robertson explains that the edge of the blade on the ice is so small that a skater's spin is about the closest thing to spinning on a vertical point as humans can do. When his body was aligned properly, Robertson says that he felt calm while spinning at his fastest, just as a top is most stable when attaining its highest speeds.

While we had the greatest spinner of all time on the phone, we couldn't resist asking him a related Imponderable: Why do almost all skating routines, in competitions and skating shows and exhibitions, end with long and fast scratch spins? Until we researched this Imponderable, we had always assumed that the practice started because skaters would have been too dizzy to continue doing anything else after rotating so fast. But Robertson pooh-poohed our theory.

The importance of the spin, to Robertson, is that unlike other spectacular skating moves, spins are sustainable. While triple jumps evoke oohs and aahs from the audience, a skater wants a spirited, prolonged reaction to the finale of his or her program. Spins are ideal because they start slowly and eventually build to a climax so fast that it cannot be appreciated without the aid of slow-motion photography.

Robertson believes that the audience remembers the ending, not the beginning, of programs. If a skater can pry a rousing standing ovation out of an audience, perhaps supposedly sober judges might be influenced by the reaction.

Robertson's trademark was not only a blindingly fast spin but a noteworthy ending. He used his free foot to stop his final spin instantly at the fastest point. Presumably, when he stopped, he opened his eyes to soak in the appreciation of the audience.

Submitted by Barbara Harris Polomé of Austin, Texas. Thanks also to David McConnaughey of Cary, North Carolina.

Why Do Straws in Drinks Sometimes Sink and Sometimes Rise to the Surface?

The movement of the straw depends upon the liquid in the glass and the composition of the straw itself. The rapidly rising straw phenomenon is usually seen in glasses containing carbonated soft drinks. Reader Richard Williams, a meteorologist at the National Weather Service, explains the phenomenon:

> . . . the rise occurs as carbon dioxide bubbles form on both the outside and inside of the straw. This increases the buoyancy of the straw and it gradually rises out of the liquid.
>
> The gas is under considerable pressure when the drink is first drawn or poured. When that pressure is released the gas forms small bubbles on the sides of the glass and on the straw. As the bubbles grow the straw becomes buoyant enough to "float" higher and higher in the container.

Occasionally, though, a straw will rise in a noncarbonated beverage, and we didn't get a good explanation for this phenomenon until we heard from Roger W. Cappello, president of strawmaker Clear Shield National. We often get asked how our sources react to being confronted with strange questions. The only answer we can give is—it varies. Sure, we like authoritative sources who fawn over us and smother us in data. But we must confess we have a special place in our hearts for folks like Cappello, who make us sweat a little before divulging their secrets. Here is his letter to *Imponderables*, verbatim, skipping only the obvious pleasantries:

> After pondering your question for a while, I decided to toss your letter as I was too busy for this. I later retrieved the letter and decided I would attempt to give you an answer that is slightly technical, mixed with some common sense and some B.S.
>
> First off, I know the action you were referring to had something to do with "specific gravity." Specific gravity, as defined by Webster, is "the rate of the density of a substance to the density

DAVID FELDMAN

of a substance (as pure water) taken as a standard when both densities are obtained by weighing in air."

Straws today are formed from polypropylene, whereas many years ago they were made of polystyrene, before that paper, and before that, wheat shafts.

Assuming water has a specific gravity of 1, polypropylene is .9, and polystyrene is 1.04. A polypropylene straw will float upward in a glass of water, whereas a polystyrene straw will sink. However, a polystyrene straw will float upward in a carbonated drink as the carbonation bubbles attach themselves to the side of the straw, which will help offset the slight specific gravity difference between water and polystyrene. A polypropylene straw will float higher in a carbonated drink for the same reason. If you put a polypropylene straw in gasoline, and please don't try this, it will sink because the specific gravity of gas is lighter than water.

If you lined up ten glasses of different liquids, all filled to the same level, the straws would most likely float at all different levels due to the different specific gravities of the liquids and the attachment of various numbers of bubbles to the straws.

I really wish you hadn't brought this up as I'm going to lunch now. I think I'll order hot coffee so I can ponder the imponderables of my business without distraction.

Good luck.

We can use all that good luck you were wishing us. I'm sure you had a productive lunch, too. Anyone willing to share information with us can eat (and sleep) with a clear conscience, knowing that he has led to the enlightenment of his fellow humans.

Submitted by Merrill Perlman of New York, New York.

Why Is the Tenor Oboe Called an "English Horn" When It Is Neither English Nor a Horn?

Dr. Kristin Thelander, professor of music at the University of Iowa School of Music, among many other experts we contacted, assured us that the "English horn" was, indeed, invented in France. No one knew exactly why or how the instrument got classified as a horn.

But the true mystery is how the credit for this instrument migrated to England. Dr. Margaret Downie Banks, curator of The Shrine to Music Museum and Center for Study of the History of Musical Instruments at the University of South Dakota, told *Imponderables* that the existence of the instrument can be traced back at least to the seventeenth century. According to Banks, in the early eighteenth century the English horn was called the *wald-hautbois* (forest oboe),

DAVID FELDMAN

a name which Baroque composers such as Johann Sebastian Bach and others italianized to *oboe da caccia* (hunting oboe).

About 1760, the name *corni inglesi* (English horn) shows up in scores for music by Haydn and Gluck; but it remains unknown just why the tenor oboe was designated the "English horn."

So what happened between the early eighteenth century and 1760 to change the name of the instrument to the English horn? Some of our experts, such as Alvin Johnson, of the American Musicological Society, and Peggy Sullivan, executive secretary of the Music Educators National Conference, were willing to speculate. They offered an oft-told but possibly apocryphal explanation: that our term is a corruption of *cor anglé*, French for "angled horn." Although they were originally straight, like "regular" oboes, instrument makers started putting angles or curves on English horns in the early eighteenth century when the instrument was used in hunting.

So, the theory goes, the English were fooled by a homonym. (*Anglé* and *anglais* do sound alike in French.) And being good chauvinists, the angled horn metamorphosed into the English horn.

We documented many instances of English words and phrases that were based on mispronunciations or misunderstandings of foreign terms in *Who Put the Butter in Butterfly?* So we'll bite on this theory.

Submitted by Robert C. Probasco of Moscow, Idaho.

DO PENGUINS HAVE KNEES? 39

Why Are Our Fingers Different Lengths? For Example, Is There a Reason Why the "Pinkie" Is Shorter Than the "Index"?

About the only angry letters we get around here are responses to answers of ours that assume the validity of evolutionary theory. But if you ask an authority, such as Dr. William P. Jollie, chairman of the Anatomy Department at the Medical College of Virginia, about this Imponderable, an evolutionary approach is what you are going to get:

> . . . anatomically, fingers are digits (our other digits are toes) and people, like all four-legged vertebrate animals, have digits characteristic both for the large group to which they belong (called a *class:* amphibians, reptiles, mammals) and for a smaller group within the class (called an *order:* rodents, carnivores, primates). So we have five fingers of a length that is characteristic for the hands of primate mammals.

DAVID FELDMAN

Of course, there is variation among different species and even variation among individual members of the same species. Some people have ring fingers noticeably longer than their index fingers; in others, the fingers are the same length. We once knew a woman whose second toe was an inch or more longer than her "big" toe.

But is there any rhyme or reason for the relative size of our digits? Dr. Duane Anderson, of the Dayton Museum of Natural History, was the only source we contacted who emphasized the role of the fingers (and hands) in grabbing objects:

> Pick up a tennis ball and you will see the fingers are all the same length. Length is an adaptation to swinging in trees initially, and then picking things up. An "even hand" would be less versatile. A long little finger, for example, would get smashed more often.

Biologist John Hertner, of Kearney State College, says that two characteristics of the digits of higher vertebrates reflect possible reasons for the unequal lengths. First, there is evidence that we can locomote more effectively with smaller outer toes. Second, over time, many higher vertebrates have a tendency to lose some structures altogether (e.g., horses have lost all but one toe).

Might humans lose a digit or two in the next few hundred million years? Unfortunately, neither the evolutionists nor the creationists will be here to find out.

Submitted by Marisa Peacock of Worcester, Massachusetts.

Why Are Sticks of Margarine and Butter Thicker and Shorter in the Western United States and Longer and Narrower in the East?

Who says we don't tackle important questions in the *Imponderables* books?

We'd love to develop a Freudian analysis to explain this phenomenon. (Are sticks of margarine phallic symbols more threatening to westerners?) Or perhaps a sociological one. (Might the fitness-crazed westerners feel superior to their stubby little western sticks?) But the real answer is a tad more prosaic.

Until recent times, dairies were local or regional in their distribution. For reasons that nobody we contacted could explain, what the industry refers to as the "western-style stick" developed out of local custom. When the behemoth dairy companies attained national distribution, they soon found that it was easier to reconfigure their molds than it was to change consumers' preferences.

So large companies like Breakstone and Land O'Lakes make two different packages, one for the West and the other for the rest of the country. In many cases, the western sticks are packaged four in a row, while the eastern counterparts are placed two by two. This also, of course, makes no particular sense.

Submitted by Alan B. Heppel of West Hollywood, California. Thanks also to Jeff Sconyers of Seattle, Washington, and Connie Krenz of Bloomer, Wisconsin.

DAVID FELDMAN

"USE OF THIS MILK CRATE AS BOOKCASE OR TV TABLE COULD BE HAZARDOUS TO YOUR HEALTH."

GEE~ AND I THOUGHT **RADON** WAS BAD!!

Why Do Plastic Milk Cases Contain a Warning That Their "Unauthorized Use Is Illegal and Enforced by Health Department and Penal Codes"?

What do certain firearms, heroin, and milk cartons have in common? Possession of each may be punishable by law. Although the wording above is used in California, many states forbid the unauthorized use of milk cartons.

Why the fuss? To get the answer, we contacted our favorite dairy Imponderables solver, Bruce Snow, recently retired from the Dairylea Cooperative in New York. If you wish to retain your faith in humanity, you may want to skip this explanation:

> For more years than I care to remember, the milk dealers in New York (and other urban areas) lost several million dollars a year on purloined plastic milk cases. Never in the history of man has anything been invented for which so many uses have been found: bookcases; sidewalk flower displays; tool chests; album

storage; toy boxes; transport cases for miscellanea; step stools, etc., *ad infinitum*.

It finally got so bad that milk dealers petitioned their legislators to make possession of a dealer's milk case illegal, subject to a fine. The dealer must have his name on the case, by law.

Theft wasn't all the dealers were contending with. Supermarkets, which daily received thousands of cases around New York, were profligate in their use of plastic cases [which currently cost $2.50 or more] to build displays, cart trash, and carry stuff home. Why not? It didn't cost them anything.

Now, however, the New York State law requires that markets must account for all cases to their milk suppliers. They are required by law to pay $2.00 for every case unaccounted for. Consequently, there is now much stricter control over the cases.

As a final note, it was discovered by some city milk dealers that shipments of new cases were being stolen, sent to plastic recyclers for some amount of money, to reappear as any one of a vast multitude of plastic gewgaws. All in all, it was a very big rathole through which a big piece of milk-generated money (ultimately from consumers) was being lost. The law has not eliminated all theft, but it has sharply reduced the problem.

The great irony, of course, is that the milk carton laws turn the tables on retailers. Supermarkets, often so reluctant to process recycled bottles and cans, now must do the same thing themselves, further proof that recycling efforts seem to work only when strong financial incentives exist.

Submitted by Mitch Hubbard of Rancho Palos Verdes,
California. Thanks also to Gregory Reis of Torrance, California.

Why Does Shampoo Lather So Much Better on the Second Application?

Even if our hands and hair are already wet, we can't seem to get a healthy lather on the first try when we shampoo our hair. But

DAVID FELDMAN

after we rinse, the shampoo foams up like crazy. Why is lather more luxuriant the second time around?

Evidently, it's because we have greasy hair, according to Dr. John E. Corbett, vice-president of technology at Clairol:

> In the first shampoo application, the lather is suppressed by the oils in the hair. When the oils are rinsed off [by the first application], the shampoo lathers much better on the second application.

Submitted by Joe Schwartz of Troy, New York.

Why Don't Cigarette Butts Burn? Is There a Particular Barrier Between the Tobacco and the Filter That Prevents the Burn?

Even cigarettes without filters don't burn quickly. If the shredded tobacco is packed tightly enough, not enough oxygen is available to feed the combustion process. The degree of porosity of the paper surrounding the tobacco rod can also regulate the degree of burn.

On a filter cigarette, however, an extra impediment is placed on the combustion process; luckily, it is not asbestos. Mary Ann Usrey, of R. J. Reynolds, explains:

> The filter is attached to the tobacco rod by a special "tipping" paper which is essentially non-porous. This paper acts to extinguish the burning coal by significantly reducing the available oxygen. So, in effect, there *is* a barrier between the tobacco and the filter, but it is *around* the cigarette, not actually between the tobacco and the filter in the interior of the cigarette.

Submitted by Frank H. Anderson of Prince George, Virginia.

What Are You Hearing When You Shake a Light Bulb?

Would you believe the ocean? We didn't think so.

Actually, what you are hearing depends upon whether you are shaking a functional or a burned-out bulb. If you are shaking a newish, functioning bulb, chances are you are hearing the delightful sound of loose tungsten particles left over in the bulb's glass envelope during its manufacturing process.

According to Peter Wulff, editor of *Home Lighting & Accessories*, these loose particles don't affect the bulb's operation or lifespan. Wulff adds that although the tungsten particles aren't left in the bulb deliberately, at one time manufacturers of high-wattage tungsten halogen bulbs did leave such residue: "Occasionally, it was recommended that after use and after the bulb cooled, the bulb should be turned upside down and then shaken to allow the loose particles to clean the inside of the glass."

But today if you hear something jangling around, chances are that you are shaking a burned-out bulb. In fact, this is the

DAVID FELDMAN

way most consumers determine whether a bulb is "dead." Richard H. Dowhan, of GTE Products, told *Imponderables* that in this case you are hearing particles of a broken filament, "the most common type of bulb failure." Barring the rare case of loose tungsten particles inside the bulb, Dowhan says "you should hear nothing when you shake a light bulb that is still capable of lighting."

Submitted by Kari Rosenthal of Bangor, Maine.

Why Do Fluorescent Lights Make a Plinking Noise When You Turn Them On?

We went to Peter Wulff again for our answer. Older fluorescent fixtures used a "preheat system," which featured a bimetallic starter (the small, round, silver piece). Wulff told us that inside the starter is a bimetallic switch which "pings" when energized. Newer fluorescent systems, such as the "preheat" or "rapid start," are rendering the "ping" a relic of our nostalgic past.

Submitted by Van Vandagriff of Ypsilanti, Michigan. Thanks also to Kathleen Russell of Grand Rapids, Michigan; Cuesta Schmidt of West New York, New Jersey; and Walter Hermanns of Racine, Wisconsin.

Why Do Cats Like So Much to Be Scratched Behind the Ears?

Most cats like to be scratched for the same reason that most humans like to be massaged: It feels good. According to veterinarian John E. Saidla, assistant director of the Cornell Feline Health Center,

> Most cats like to have their total bodies rubbed or stroked by humans. A cat's skin is chock full of nerve endings, making your stroking a sensual experience.

But *our* skin is full of nerve endings, and not too many of us start wiggling our legs with delight when we're scratched behind the ears. But then again, unlike cats, we don't tend to have ear mites. Dr. Saidla explains:

> Most cats harbor very few mites, while others have huge infections that are causing serious clinical problems. The mite in the

DAVID FELDMAN

ear canal burrows into the layers of skin lining the ear canal. The cat is allergic, or at least, reacts to the feces and enzymes the mite produces, resulting in pruritus or itching. When the owner rubs the skin behind the ear, it feels good and the cat responds appreciatively.

Submitted by Robert J. Abrams of Boston, Massachusetts.

Why Aren't There Plums in Plum Pudding? And Why Is It Called a Pudding Rather Than a Cake?

Even though it contains flour and is as sweet and rich as any cake, plum pudding cannot be classified as a cake because it contains no leavening and is not baked, but steamed.

Besides flour, plum pudding contains suet, sugar, and spices and is studded with raisins and currants. In early America, both raisins and currants were referred to as "plums" or "plumbs." And presumably because the raisins and currants were the only visually identifiable ingredients in the dessert (which traditionally was served after the pumpkin pie at Thanksgiving), the nickname stuck.

Come to think of it, plum pudding isn't the only weirdly named dessert served at Thanksgiving. We used to ask our parents where the meat was in mincemeat pie.* Give us an honest pumpkin pie any day.

Submitted by Bert Garwood of Grand Forks, North Dakota.

---

* To forestall a flood of letters, may we say on the record that mincemeat pie may or may not have minced meat in it.

What Do the Little Red Letter and Number Stamped on the Back of My Envelope Mean?

When the arm of the United States Postal System's Multi-Position Letter Sort Machine (or, as we close friends like to call it, MPLSM) picks up an envelope, it automatically stamps a letter and number on the back. The letter identifies which MPLSM processes the piece, and the number singles out which console on the machine, and thus which MPLSM operator, handles the piece.

This code has nothing to do with delivering the letter. The code is simply a way for the USPS to identify when a particular machine, or its operator, is malfunctioning. In other words, the A4 on the back of your envelope is the equivalent of the "Inspected by 8" label you sometimes find in the pocket of your new jacket.

Although it is our experience that most of the MPLSM codes are red, a casual glance at our voluminous mail indicates that brown and purple are popular, too. Frank P. Brennan, Jr., general manager of media relations for the USPS, says that each tour or shift has its own color.

Not every letter is processed by a MPLSM; increasingly, the USPS is relying on optical scanners. Scanners may be faster than MPLSMs, but without that red code, they can't be held as accountable as MPLSMs when they screw up, either.

DAVID FELDMAN

Why Do Owners or Handlers Use the Word "Sic" to Instruct a Dog to "Get Him"?

Dog World magazine was kind enough to print our query about this Imponderable in their June 1990 issue. We were soon inundated with letters from dog lovers, the most comprehensive of which came from Fred Lanting, of the German Shepherd Dog Club of America:

> The command "sic" comes from a corruption of the German word *such*, which means to seek or search. It is used by Schutzhund [guardian and protection] and police trainers as well as by people training dogs for tracking. If the command "sic" is issued, it means that the dog is to find the hidden perpetrator or victim. In German, *sic* is pronounced "sook" or "suk," but like many foreign words, the pronunciation has been altered over time by those not familiar with the language.
>
> "Sic" has developed from [what was originally] a command to find a hidden bad guy, who [in training exercises] is usually covered by a box or hiding in an open pyramidal canvas blind. Because in police and Schutzhund training the bad guy is attacked if he tries to hit the dog or run away, the word has become associated with a command to attack.

Lanting's answer brings up another Imponderable: If "sic" is a misspelling of the German word, should it be printed as " 'sic' (*sic*)"?

Submitted by Annie Lloyd of Merced, California.

In Baseball Scoring, Why Is the Letter "K" Chosen to Designate a Strikeout?

Lloyd Johnson, ex–executive director of the Society for American Baseball Research, led us to the earliest written source for this story, *Beadle's Dime Base-Ball Player*, a manual published in 1867 that explained how to set up a baseball club. Included in *Beadle's* are such quaint by-laws as "Any member who shall use profane language, either at a meeting of the club, or during field exercise, shall be fined _____cents."

A chapter on scoring, written by Henry Chadwick, assigns meaning to ten letters:

A for first base
B for second base
C for third base
H for home base
F for catch on the fly
D for catch on the bound
L for foul balls
T for tips
K for struck out
R for run out between bases

Chadwick advocated doubling up these letters to describe more events:

H R for home runs
L F for foul ball on the fly
T F for tip on the fly
T D for tip on the bound

He recognized the difficulty in remembering some of these abbreviations and attempted to explain the logic:

> The above, at first sight, would appear to be a complicated alphabet to remember, but when the key is applied it will be at once seen that a boy could easily impress it on his memory in a few minutes. The explanation is simply this—we use the first

DAVID FELDMAN

letter in the words, Home, Fly, and Tip and the last in Bound, Foul, and Struck, and the first three letters of the alphabet for the first three bases.

We can understand why the last letters in "Bound" and "Foul" were chosen—the first letters of each were already assigned a different meaning—but we can't figure out why "S" couldn't have stood for struck out.

Some baseball sources have indicated that the "S" was already "taken" by the sacrifice, but we have no evidence to confirm that sacrifices were noted in baseball scoring as far back as the 1860s.

Submitted by Darin Marrs of Keller, Texas.

What Are the Skins of Hot Dogs Made Of?

Our correspondent wondered whether hot dog skins are made out of the same animal innards used to case other sausages. We recollect when we sometimes used to need a knife to pierce a hot dog. Don't hot dog skins seem a lot more malleable than they used to be?

Evidently, while we were busy chomping franks down, manufacturers were gradually eliminating hot dog skins. Very few mass-marketed hot dogs have skins at all any more. Thomas L. Ruble, of cold-cut giant Oscar Mayer, explains:

> A cellulose casing is used to give shape to our hot dogs and turkey franks [Oscar Mayer owns Louis Rich] during cooking and smoking, but it is removed before the links are packaged. What may have seemed like a casing to you would have been the exterior part of the link that is firmer than the interior. This texture of the exterior of a link could be compared to the crust on a cake that forms during baking.

Submitted by Ted Goodwin of Orlando, Florida.

DAVID FELDMAN

Why Is Comic Strip Print in Capital Letters?

The cartoonists we contacted, including our illustrious (pun intended) Kassie Schwan, concurred that it is easier to write in all caps. We've been printing since the first grade ourselves and haven't found using small letters too much of a challenge, but cartoonists have to worry about stuff that never worries us. Using all caps, cartoonists can allocate their space requirements more easily. Small letters not only vary in height but a few have a nasty habit of swooping below or above most of the other letters (l's make a's look like midgets; and p's and q's dive below most letters).

More importantly, all caps are easier to read. Mark Johnson, archivist for King Features, reminded us that comic strips are reduced in some newspapers and small print tends to "blob up."

We wish that our books were set in all caps. It would automatically rid us of those pesky capitalization problems. While we're musing ... we wonder how *Classics Illustrated* would handle the type if it decided to publish a comics' treatment of e. e. cummings' poetry?

Submitted by Carl Middleman of St. Louis, Missouri.

Why Are Peanuts Listed Under the Ingredients of "Plain" M&Ms?

We've always felt that "peanut" M&Ms weren't as good as "plain" ones—that the synergy between Messrs. Goober and Cocoa just wasn't there. As lovers of chocolate and peanuts and, come to think of it, hard candy shells, as well, you could have knocked us over with an M&M when two readers brought it to our attention that "plain" M&Ms contain peanuts.

We contacted the folks at M&M/MARS to solve this troubling Imponderable and we heard from Donna Ditmars in the consumer affairs division. She told us that peanuts are finely ground and added to the chocolate for flavor. The quantity of peanuts in the candy is so small that "labeling laws do not require that we list this small amount of peanuts as an ingredient, we do so voluntarily so that consumers will know that it is in the candy."

Why is listing the peanuts so important? Nuts are the source of one of the most common food allergies.

FLASH: Just after the publication of the hardcover edition of *Do Penguins Have Knees?*, we heard surprising news from the external relations director of M&M/MARS, Hans S. Fiuczynski. Starting in January, 1992, the company no longer includes any peanuts in its plain candies. Even so, this Imponderable will not become obsolete. M&M/MARS will continue to list peanuts as an ingredient on the label, just in case a small amount of peanuts inadvertently appears in the plain candies.

Submitted by Martha Claiborne of Anchorage, Kentucky.
Thanks to Susan Wheeler of Jacksonville, North Carolina.

Why Do the Volume Levels of Different Cable Networks Vary Enormously Compared to Those of Broadcast TV Networks?

Anyone with an itchy hand and a remote control device that can control volume levels knows how often one must adjust the volume control when flipping around stations. Anyone without a volume control on a remote control device has probably walked the equivalent of 892 miles in round trips from the La-Z-Boy to the TV set to keep the Pocket Fisherman commercial from blasting innocent eardrums.

We were confronted with a lot of shilly-shallying about this Imponderable from folks in the cable television industry until we heard from Ned L. Mountain, chairman of the subcommittee on Quality Sound in Cable Television of the National Cable Television Association. Mountain doesn't offer any quick solutions, but he does explain the historical and technological problems involved. He lists three advantages that broadcast stations have over cable outlets in transmitting even audio levels:

> 1. The FCC mandates strict standards for broadcast maximum peak audio levels.
>
>> Most TV stations also employ sophisticated and expensive audio processing equipment to maintain a consistent level for their station. Since they only have one channel to worry about, they can afford it.
>
> 2. At the point where signals originate for a community (called the "head ends"), broadcast stations have personnel to monitor the levels. Most cable head ends don't. Humans can make "subjective audio level control" adjustments as necessary.
>
> 3. Most broadcast stations have converted to stereo; while they did so, they took the opportunity to upgrade their audio facilities. The result: a more uniform sound among broadcast outlets.

Cable operators must also conform to the FCC standard for

peak program levels, but are under numerous handicaps, as Mountain explains:

> A cable operator must attempt to achieve this standard on as many as 30 to 40 channels simultaneously from sources over which he has no control.
>
> The problems start at the sources. Most cable programs are satellite delivered using various technologies where there are no standards, only "understandings." The programmers themselves may or may not use the same type of audio processing as that employed by over-the-air broadcasters.
>
> The cable operator many times compounds this problem by inserting locally generated commercials on these channels. The sound of these may or may not match the network on any given day.
>
> This equipment is automated, and the head end is generally unattended. Without standards or expensive "automatic gain control devices, the levels can and do vary from channel to channel."

Mountain's subcommittee consists of programmers, satellite transmission experts, cable operators, and equipment manufacturers. Their first task is to get the programmers and satellite delivery systems to coordinate the quantification of their audio standards, so that operators at least know how to set levels.

Even though this problem annoys us no end, we can't help being won over by Mountain's sincerity and a hook that is going to compel every reader of this book to buy the *next* volume of *Imponderables:*

> David, I wish there were an easy answer to your question. I can assure your readers, however, that there are significant efforts being expended within our industry to investigate and solve these problems. Perhaps by the time your next issue is in print, we will have made significant inroads and I can give you an update.

We'll be waiting.

Submitted by M. Ian Silbergleid of Northport, New York.

Why Are Men's Shoe Heels Built in Layers?

Rubber is most durable and attractive material for heels, but according to Lloyd E. Brunkhorst, vice-president of research and engineering for Brown Shoe Company,

> A thick rubber heel is often too soft, and the result is instability. Therefore, a heel base made of polyethylene ½" to ¾" thick, with a ½" or so rubber top lift, gives the best performance.

William Kelly, of Brockton Sole and Plastics, echoes Brunkhorst's sentiments, and adds that polyethylene better withstands the moisture to which heels are constantly subjected.

Both sources indicated that the "stacked leather look," which is now in fashion, requires many layers of quarter-inch-thick leather to form the heel. The few shoes that are manufactured with unlayered rubber are considered to be "low class."

Polyethylene heels are cheaper than rubber, which reduces not only the cost of the shoe to the manufacturer and consumer but the cost of repairing the shoe. In many cases, the entire heel need not be replaced when damage occurs. As Brunkhorst mentions, replacement of a top lift rather than a whole heel is much less expensive, and the lifts are more readily available than whole heels.

Submitted by an anonymous caller on "The Ray Briem Show," KABC-AM, Los Angeles, California.

Speech bubble: WHOOPS! WE RAN OUT OF HANDS!!

Why Are Horses' Heights Measured to the Shoulder Rather Than to the Top of the Head?

David M. Moore, Virginia Tech University's veterinarian and director of the Office of Animal Resources, compares measuring a horse to trying to measure a squirming child. At least you can back a child up to a wall. If the child's legs, back, and neck are straight, the measurement will be reasonably accurate:

> But with a horse, whose spinal column is parallel to the ground (rather than perpendicular, as with humans), there is no simple way to assure that each horse will hold its head and neck at the same point. Thus, measurements to the top of the head are too variable and of little use.

Dr. Wayne O. Kester, of the American Association of Equine Practitioners, told *Imponderables* that when a horse is standing squarely on all four feet, the top of the withers (the highest point

DAVID FELDMAN

on the backbone above the shoulder) is always the same *fixed* distance above the ground, thus providing a consistent measurement for height. Kester estimates that "head counts" could vary as much as two to six feet.

Submitted by Gavin Sullivan of Littleton, Colorado.

Why Are the Edges on the Long Side of Lasagna Usually Crimped?

Farook Taufiq, vice-president of quality assurance at The Prince Company, had no problem answering this Imponderable:

> The curls at the edge of lasagna strips help retain the sauce and the filling between layers. If the lasagna strips are flat, the sauce and the filling will slip out from between layers while cooking as well as while eating.

Now if someone will only invent a method of keeping lasagna (and its sauce) on our fork while it makes the arduous journey from the plate to our mouths, we would be most appreciative.

Submitted by Sarah Duncan of Mars, Pennsylvania.

What Happens to Your Social Security Number When You Die? How and When, If Ever, Is It Reassigned?

You don't need to be a hall-of-famer to get this number retired. John Clark, regional public affairs officer of the Social Security Administration, explains:

Each number remains as unique as the individual it was first assigned to. When someone dies, we retire the number.

The first number was issued in 1936. The nine-digit system has a capacity for creating nearly one billion possible combinations. A little more than a third of the possible combinations have been issued in the fifty-five years since the first number was issued.

It's comforting to know that you can take *something* with you.

Submitted by Albert Mantei of Crystal, Minnesota.

"I love your cologne... what is it?"

it's **new car**

new car *femmes*
for women...

new car
...men

new car $
voitures
... old cars

that musky, vinyl-y fragrance everyone **loves!!!**

What Exactly Are We Smelling When We Enjoy the "New-Car Smell"?

You didn't think that only one ingredient could provide such a symphony of smells, did you? No. Detroit endeavors to provide the proper blend of constituents that will provide you with the utmost in olfactory satisfaction. (We won't even talk about the exotic scents of European and Asian cars.) C. R. Cheney, of Chrysler Motors, provided us with the most comprehensive explanation and the most poignant appreciation:

> The smell we all enjoy inside a new vehicle (that "new-car smell") is a combination of aromas generated by fresh primer and paint, and the plastic materials used on instrument panels, around the windows, and on door trim panels. Plus, there are odors given off by carpeting, new fabrics, leather, and vinyl used for soft trim and upholstery. Rubber, adhesives, and sealers also play a part in creating this unique smell that never lasts as long as we would like and seems nearly impossible to duplicate.

Submitted by William Janna of Memphis, Tennessee. Thanks also to Jerry Arvesen of Bloomington, Indiana, and David Nesper of Logansport, Indiana.

DO PENGUINS HAVE KNEES?

Why Are Some Cleansers Marked "For Industrial or Commercial Use Only"? How Are They Different from Household Cleansers?

With few exceptions, the chemicals and detergents used in commercial cleansers are no different from those marketed to home consumers, although industrial-strength cleanser is likely to contain much less water than Fantastik or Mr. Clean. Why? The answer has more to do with marketing and sociology than technology.

Until the 1950s, most cleaning was done with soaps (fatty acids and lye) rather than detergents (made from alkaline substances). Unlike detergents, soap didn't need much water to add to its cleaning effectiveness. When using soap, consumers rarely added water.

When synthetic detergents were introduced in the 1950s, most home consumers didn't adjust properly. According to Tom Mancini, of U.S. Polychemical, manufacturers were forced to add water to detergents designed for home use because consumers wouldn't add enough water to the products to make them work effectively. Consumers also enjoyed the convenience of applying cleanser directly to a sponge or dirty surface rather than first diluting the detergent with water.

Of course, consumers have had to pay for the privilege; commercial cleansers are much cheaper than home equivalents, and not just because industry buys cleansers in bulk. When you buy a cleanser in a big plastic package at the supermarket, you are carrying mostly a big package of water, the equivalent of buying a package of ready-to-drink ice tea rather than a jar of iced tea mix.

Industrial users have totally different priorities. They are quite willing to sacrifice a little convenience to save money; by buying a concentrated product, companies can save on unnecessary packaging. Professional cleaners also realize that detergents need to be diluted to work effectively. In almost all cases,

DAVID FELDMAN

"industrial-strength" cleansers can be used in the home if diluted sufficiently.

There is one major difference in the ingredients of home and industrial cleansers. Home consumers care about how their cleansers smell. In most cases, corporate decision makers don't care much about the smell of cleansers (although the janitors, who have to work with the stuff all the time, undoubtedly do); as a result, many household cleansers contain perfume to mask the odor of unpleasant ingredients. Perfume jacks up the price of the product without adding anything to its cleaning ability.

Submitted by Jeffrey Chavez of Torrance, California.

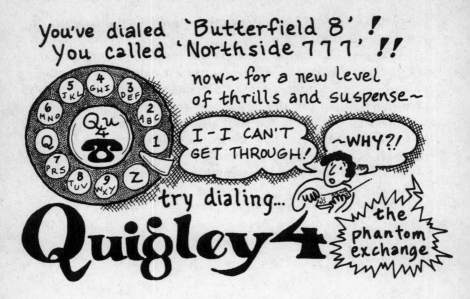

Why Are the Letters "Q" and "Z" Missing from the Telephone Buttons?

The whereabouts of the missing "Q" and Z" are very much on the minds of *Imponderables* readers. In fact, this is easily one of our top ten most frequently asked questions. We have heretofore restrained ourselves from answering it because we've seen the solution bandied about in print already. So we won't call it an Imponderable (we called these questions that have already been in print but just won't go away Unimponderables in *Why do Dogs Have Wet Noses?*), but we will answer it anyway, since it segues neatly into the next Imponderable, a phenomenon less often written about.

Before the days of area codes, operators connected all long distance calls and many toll calls. When the Bell system started manufacturing telephones with dials, users were able to make many of their own local and toll connections. When direct dial-

DAVID FELDMAN

ing was instituted, phone numbers consisted of two letters and five numbers. A number we now call 555-5555 might have then been expressed as KL5-5555. And the phone company provided a nifty mnemonic for each exchange.

So the phone company assigned three letters, in alphabetical order, to each dial number. The number one was skipped because one was assigned as an access code and for internal phone company use (many phone companies used three-digit numbers starting with 1-1 . . . for directory assistance and repair lines); the zero was avoided because it automatically summoned the operator, regardless of subsequent numbers dialed. So there were eight numbers on the dial that needed letters and twenty-six letters available. Eight goes into twenty-six an inconvenient three and one-quarter times. Two of the letters had to be discarded.

Sure, the phone company could have simply dropped the last two letters of the alphabet, but in our opinion they selected well. What letters are less commonly used and more easily discarded than the two letters valuable only to Scrabble players—"Q" and "Z"? "Q" would have been a problematic choice at best. How can you make an effective mnemonic when virtually all words starting with "Q" must be followed by a "U"? If "Q" had its "rightful" place on the number 7, 8 (where "U" is located") would usually have to follow, severely limiting the numbers assignable to the exchange.

"Z," of course, is the last letter and accustomed to suffering the usual indignities of alphabetical order. Maybe the thought of a phone number starting with "ZEbra," "ZInnia," or "ZAire" is overwhelmingly exciting to someone, but for the most part its loss has not been missed.

Submitted by Robert Abrams of Boston, Massachusetts, and a cast of thousands.

Why Is the Middle Digit of North American Area Codes Always a 0 or a 1?

The Bell system introduced three-digit area codes in 1945. Bell was quite aware of the cost savings in direct dialing for long distance calls but also knew that unless it could devise a system to distinguish area codes from the first three digits of ordinary local phone numbers, an operator would have to switch calls.

All ten-digit phone numbers consist of three parts: an area code (the first three numbers); an office code (the next three numbers); and a line number (the last four numbers). We have already explained in the last Imponderable why no office code could start with a 0 or 1. When the Bell system created the area code, it initially extended the "ban" on zeroes and ones to the second digit of the office code as well. By assigning all area codes a second digit of either 0 or 1, automatic switching equipment could differentiate between long distance calls and local or toll calls and route them accordingly. The equipment could also sort calls by the first digit—if the initial digit is a 1, a ten-digit number will follow; while an initial 2–9 means a seven-digit number will follow.

When the area code system was first instituted, all states with only one area code had a 0 as the middle digit; states with more than one area code used 1 for the middle digit of the area code. This practice had to be abandoned when the Bell system ran out of ones as more states needed more than one area code. Now, some populous states have area codes with middle digits of 0.

Because people and telephones have proliferated, the numbering system has had to change several times. The original configuration of office codes yielded a limit of 640 different numbers. To increase the number of office codes available, zeroes and ones have been added to the second digit of office codes, allowing for an eventual expansion of 152 extra office codes.

By the twenty-first century, we would probably run out of

DAVID FELDMAN

area codes if we kept the same numbering method. The phone system is preparing to introduce middle digits other than one or zero in the next century. As long as all long distance calls are preceded by one, it won't be a problem.

Submitted by Carol Oppenheim of Owings Mill, Maryland.
Thanks also to Nicole Donovan of Wenham, Massachusetts.

Why Were Duels Always Fought at Dawn? Or Is This Depiction in Fiction and Movies Not True?

Not true, we're afraid. Historians assured *Imponderables* that duels were fought at any time of the day. But dawn was definitely the preferred time; a duel fought in twilight could turn into more of a crapshoot than a gunshoot.

Doesn't make much sense to us. We might be convinced to get up at dawn to go fishing. But if we knew we had an approximately fifty-fifty chance of dying on a particular day, we'd at least want a decent night's sleep the night before and time for a doughnut or two before we fought.

Historian C. F. "Charley" Eckhardt speculates on this strange predilection of duelists to fight to the death at inconvenient hours:

> Just at sunrise, if the list [the technical term for a dueling ground] was oriented north-south, neither man got the sun-to-the-back advantage. Also, either the local law was still abed or, if there was a regular police force in the area, the day watch and night watch were changing shifts. Fighting at dawn minimized the likelihood of interference by the law, the same reason why many burglaries occur between 3 and 4 P.M. and 11 P.M. to midnight. Most police departments change shifts at 3 P.M., 11 P.M., and 7 A.M.

Submitted by Jan Anthony Verlaan of Pensacola, Florida.

Quaintly Earnest
LIBERAL ARTS COLLEGE
no tuition · no requirements
f. 1968

What Exactly Are the Liberal Arts, and Who Designated Them So?

Our correspondent, Bill Elmendorf, contacted two four-year colleges and one two-year college for the answer to this question. Despite the fact that they were liberal arts colleges, none of the officials he spoke to could answer this question. Evidently, a good liberal arts education doesn't provide you with the answer to what a liberal art is.

Actually, a consultation with an encyclopedia will tell you that the concept of the liberal arts, as developed in the Middle Ages, involved seven subjects: grammar, logic, rhetoric, arithmetic, geometry, music, and astronomy. Why astronomy and not biology? Why rhetoric and not art? For the answers to this question, we have to delve into the history of the liberal arts.

Our expression is derived from the Latin *artes liberalis*, "pertaining to a free man." Liberal arts are contrasted with the "servile" arts, which have practical applications. As educator

DAVID FELDMAN

Tim Fitzgerald wrote *Imponderables,* "the liberal arts were considered 'liberating,' enabling the student to develop his or her potential beyond the mundane, to create, to be fully human, to (in the medieval mindset) believe."

The notion of seven ennobling arts emerged long before the Middle Ages. In Proverbs 9:1, the Bible says, "Wisdom hath builded her house, she hath hewn out her seven pillars." Robert E. Potter, professor of education at the University of Hawaii at Manoa, wrote *Imponderables* a fascinating letter tracing the history of the liberal arts. Before the birth of Christ and into the first century A.D., Roman writers like Cicero and Quintilian discussed the proper curriculum for the orator and public leader. Varro (116–27 B.C.) listed in his *Libri Novem Disciplinarum* the seven liberal arts but also included medicine and architecture.

Potter mentions that in the early Christian era, church elders opposed the classical liberal arts. Perhaps the most stirring condemnation was written in the Apostolic Constitutions in the third century:

> Refrain from all the writings of the heathen for what has thou to do with strange discourses, laws, or false prophets, which in truth turn aside from the faith for those who are weak in understanding? For if thou wilt explore history, thou hast the Books of the Kings; or seekest thou for words of wisdom and eloquence, thou hast the Prophets, Job, and the Book of Proverbs, wherein thou shalt find a more perfect knowledge of all eloquence and wisdom, for they are the voice of the Lord.

Later Christian scholars, including Augustine, embraced the study of the liberal arts.

Potter calls Martianus Capella of Carthage's *The Marriage of Philology and Mercury* the "definitive" work on the liberal arts:

> This fourth-century allegory had nine books. The first two described the wedding of the daughter of Wisdom, a mortal maiden who represented schooling, and Mercury, who, as the inventor of letters, symbolized the arts of Greece. The remaining seven books describe the bridesmaids. Apollo did not admit two

other "bridesmaids," medicine and architecture, "inasmuch as they are concerned with perishable earthly things."

Many people attack the modern liberal arts education, saying that little is taught that pertains to our actual lives now. Little do they know that this lack of "relevance" is precisely what characterized the liberal arts from their inception. In ancient times, servile folks had to sully themselves with practical matters like architecture, engineering, or law. Only the elite freemen could ascend to the lofty plateau of the contemplation of arithmetic.

Today, the meaning of liberal arts is murky, indeed. Art, other hard sciences besides astronomy, foreign languages, philosophy, history, and most social sciences are often included under the umbrella of liberal arts. Just about any school that *doesn't* train you for a particular profession is called a liberal arts institution.

Submitted by Bill Elmendorf of Lebanon, Illinois. Thanks also to Brianna Liu of Minneapolis, Minnesota.

Why Do Birds Tend to Stand on One Foot While Sleeping? Why Do Birds Tend to Bury Their Heads Under Their Wings While Sleeping?

In *When Do Fish Sleep?*, we discussed the amazing locking mechanism of birds' toes that enables them to perch on telephone wires without falling off. In fact, they can perch just as easily while standing on only one leg. Since they can balance as easily on one leg as two, one of the main reasons for perching on one leg (whether or not they are sleeping) is simply to give the other leg a rest.

But birds also seek warmth, and perching on one foot gives them a "leg up" on the situation, as Nancy Martin, naturalist at the Vermont Institute of Natural Science, explains:

DAVID FELDMAN

Since birds' feet are not covered with feathers, they can lose significant amounts of body heat through their feet, especially when standing on ice or in cold water. With their high metabolic rates, birds usually try to conserve as much energy as possible, hence the habit of standing on one leg.

A corollary: Birds also stick their head under their feathers to preserve heat.

Submitted by Lee Dresser of Overland Park, Kansas. Thanks also to Jocelyn Noda of Los Angeles, California.

Why Is a Marshal or Sheriff's Badge Traditionally a Five-Pointed Star but a Deputy's Six-Pointed?

The five-pointed pentacle is the symbol of the United States Marshal's Service. In ancient times, the pentacle was used by sorcerers and believed to impart magical powers. As late as the sixteenth century, soldiers wore pentacles around their necks in the belief that they made them invulnerable to enemy missiles.

But it turns out that even early American lawmen forged a new tradition of forsaking old traditions at the drop of a hat. It just isn't true that sheriffs always wore five-pointed stars and their deputies six-pointed ones. Charles E. Hanson, Jr., director of The Museum of the Fur Trade in Chadron, Nebraska, wrote *Imponderables* that one could despair of trying to find logic to the patterns of badges:

> There seems to be no fixed protocol on five- and six-pointed badges. In America, the five-point star has been preeminent from the beginning. It is the star in the flag, in the insignia of an army general, and on the Medal of Honor. It was obviously the logical choice for the first sheriffs' badges.
>
> When other shapes began to be used for badges, it seemed right that circles, shields, and six-pointed stars would be used for lesser legal representatives than the top lawman.

DO PENGUINS HAVE KNEES? 73

This didn't hold true indefinitely. Our library has a 1913 supply catalog which offers five-point stars engraved "City Marshal" or "Chief of Police" and six-point stars engraved, "City Marshal," "Sheriff," "Constable," "Detective," etc.

Historian Charley Eckhardt has even developed a theory to explain why the five-point might have been inflated to six points: It was simply too hard to make a five-pointed star.

> The five- and six-pointed star "tradition" seems to be purely a twentieth-century one. I've seen hundreds of badges from the nineteenth century, and they ranged from the traditional policeman's shield to a nine-pointed sunburst. Five- and six-pointed stars predominated, but in no particular order—there was no definite plurality of five points in one group and six points in another. I have noticed, however, that the majority of the *locally* made star-shaped badges produced outside of Texas were six-pointed. There may be a reason for that.
>
> When you cut a circle, if you take six chords equal to the radius of the circle and join them around the diameter, you will find that the chords form a perfect hexagon. If you join alternate points of the hexagon, you get two superimposed equilateral triangles—a six-pointed star. In order to lay out a pentagon within a circle—the basic figure for cutting a five-pointed star—you have to divide the circle into 72-degree arcs. This requires a device to measure angles from the center—or a very fine eye and a lot of trial and error. Since many badges, including many deputy sheriff and marshal badges, were locally made, it would have been much easier for the blacksmith or gunsmith turned badgemaker for a day to make a six-pointed star.

Who says that the shortage of protractors in the Old West didn't have a major influence on American history?

Submitted by Eugene S. Mitchell of Wayne, New Jersey. Thanks also to Christopher Valeri of East Northport, New York.

Foundation Sale!
"Mystery In Lace"

When you put on our girdle...

(you'll never want to peel it off!)

...thousands of delicate lace flowers absorb and hold your excess er~FAT!

#29⁹⁰

When You Wear a Girdle, Where Does the Fat Go?

Depends upon the girdle. And depends upon the woman wearing the girdle.

Ray Tricarico, of Playtex Apparel, told *Imponderables* that most girdles have panels on the front to help contour the stomach. Many provide figure "guidance" for the hips and derriere as well.

"But where does the fat go?" we pleaded. If we cinch a belt too tight, the belly and love handles plop over the belt. If we poke ourselves in the ribs, extra flesh surrounds our fingers. And when Victorian ladies wore corsets, their nineteen-inch waists were achieved only by inflating the hips and midriff with displaced flesh. Mesmerized by our analogies, Tricarico suggested we contact Robert K. Niddrie, vice-president of merchandising at Playtex's technical research and development group. We soon discovered that we had a lot to learn about girdles.

First of all, "girdles" may be the technical name for these

undergarments, but the trade prefers the term "shapewear." Why? Because girdles conjure up an old-fashioned image of undergarments that were confining and uncomfortable. Old girdles had no give in them, so, like too-tight belts, they used to send flesh creeping out from under the elastic bands (usually under the bottom or above the waist).

The purpose of modern shapewear isn't so much to press in the flesh as to distribute it evenly and change the contour of the body. And girdles come in so many variations now. If women have a problem with fat bulging under the legs of the girdle, they can buy a long-leg girdle. If fat is sneaking out the midriff, a high-waist girdle will solve the problem.

Niddrie explains that the flesh is so loose that it can be redistributed without discomfort. Shapewear is made of softer and more giving fabrics. The modern girdle acts more like a back brace or an athletic supporter—providing support can actually feel good.

"Full-figured" women are aware of so-called minimizer bras that work by redistributing tissue over a wider circumference. When the flesh is spread out over a wider surface area, it actually appears to be smaller in bulk. Modern girdles work the same way. You can demonstrate the principle yourself. Instead of poking a fatty part of your body with your finger, press it in gently with your whole hand—there should be much less displacement of flesh.

Niddrie credits DuPont's Lycra with helping to make girdles acceptable to younger women today. So we talked to Susan Habacivch, a marketing specialist at DuPont, who, unremarkably, agreed that adding 15 to 30 percent Lycra to traditional materials has helped make girdles much more comfortable. The "miracle" of Lycra is that it conforms to the body shape of the wearer, enabling foundation garments to even and smooth out flesh without compressing it. The result: no lumps or bumps. Girdles with Lycra don't eliminate the fat but they "share the wealth" with adjoining areas.

Submitted by Cynthia Crossen of Brooklyn, New York.

DAVID FELDMAN

What Do Mosquitoes Do During the Day? And Where Do They Go?

At any hour of the day, somewhere in the world, a mosquito is biting someone. There are so many different species of mosquitoes, and so much variation in the habits among different species, it is hard to generalize. Some mosquitoes, particularly those that live in forests, are diurnal. But most of the mosquitoes in North America are active at night, and classified as either nocturnal or crepuscular (tending to be active at the twilight hours of the morning and/or evening).

Most mosquitoes concentrate all of their activities into a short period of the day or evening, usually in one to two hours. If they bite at night, mosquitoes will usually eat, mate, and lay eggs then, too. Usually, nocturnal and crepuscular female mosquitoes are sedentary, whether they are converting the lipids of blood into eggs or merely waiting to go on a nectar-seeking expedition to provide energy. Although they may take off once or twice a day to find some nectar, a week or more may pass between blood meals.

If the climatic conditions stay constant, mosquitoes tend to stick to the same resting patterns every day. But according to Charles Schaefer, director of the Mosquito Control Research Laboratory at the University of California at Berkeley, the activity pattern of mosquitoes can be radically changed by many factors:

1. Light (Most nocturnal and crepuscular mosquitoes do not like to take flight if they have to confront direct sunlight. Conversely, in homes, some otherwise nocturnal mosquitoes will be active during daylight hours if the house is dark.)
2. Humidity (Most will be relatively inactive when the humidity is low.)
3. Temperature (They don't like to fly in hot weather.)
4. Wind (Mosquitoes are sensitive to wind, and will usually not take flight if the wind is more than 10 mph.)

Where are nocturnal mosquitoes hiding during the day? Most never fly far from their breeding grounds. Most settle into vegetation. Grass is a particular favorite. But others rest on trees; their coloring provides excellent camouflage to protect them against predators.

A common variety of mosquito in North America, the anopheles, often seeks shelter. Homes and barns are favorite targets, but a bridge or tunnel will do in a pinch. Nocturnal or crepuscular mosquitoes are quite content to rest on a wall in a house, until there is too much light in the room. In the wild, shelter-seeking mosquitoes will reside in caves or trees.

We asked Dr. Schaefer, who supplied us with much of the background information for this Imponderable, whether mosquitoes were resting or sleeping during their twenty-two hours or so of inactivity. He replied that no one really knows for sure. We're reserving *When Do Mosquitoes Sleep?* as a possible title for a future volume of *Imponderables*.

Submitted by Jennifer Martz of Perkiomenville, Pennsylvania. Thanks also to Ronald C. Semone of Washington, D.C.

What Does the "CAR-RT SORT" Printed Next to the Address on Envelopes Mean?

Reader Jeff Bennett writes: "CAR-RT SORT is printed on a lot of the letters I receive. Obviously it's got something to do with mail sorting, but what does it stand for?"

Not unlike us, Jeff, it sounds like you've been receiving more than your share of junk mail lately. CAR-RT SORT is short for Carrier Route Presort and is a special class of mail. As the name implies, to qualify for the Carrier Route First-Class Mail rate, mailers must arrange all letters so that they can be given to the appropriate mail carrier without any sorting by the postal system. Each piece must be part of a minimum of ten pieces for that carrier; if there are not ten pieces for a particular carrier route, the mailer must pay the rate for Presorted First-Class Mail. (Presorted First-Class Mail costs more than Carrier Route because it requires only that the mail be arranged in ascending order of ZIP code.)

Don't expect to see CAR-RT SORT on the envelopes of your friends' Christmas cards. Carrier Route Mail must be sent in a single mailing of not less than 500 pieces. And if they really have ten friends serviced by one postal carrier, it would be cheaper for them to hand-deliver the cards.

Submitted by Jeff Bennett of Poland, New York. Thanks also to Matt Menentowski of Spring, Texas.

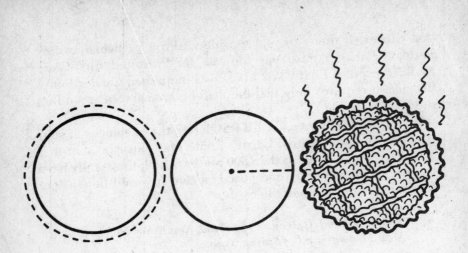

Why Was "pi" Chosen as the Greek Letter to Signify the Ratio of a Circle's Circumference to Its Diameter?

The history of "pi" is so complex and fascinating that whole books have been written about the subject. Still, if you want to make a long story short, it can be boiled down to the explanation provided by Roger Pinkham, a mathematician at the Stevens Institute of Technology in Hoboken, New Jersey:

> There are two Greek words for perimeter, *perimetros* and *periphireia*. The circumference of a circle is its perimeter, and the first letter of the Greek prefix *peri* (meaning "round") used in those two words was chosen.

Mathematicians attempted to calculate the pi ratio before the birth of Christ. But it wasn't the ancient Egyptians or Greek mathematicians who first coined the term. San Antonio math teacher John Veltman sent us documentation indicating that although pi was earlier applied as an abbreviation of "periphery," the first time it was found in print to express the circle ratio was in 1706, when an English writer named William Jones, best known as a translator of Isaac Newton, published *A New Introduction to Mathematics*. "Pi" did not enjoy widespread accep-

DAVID FELDMAN

tance until 1737, when Jones's term was popularized by the great Swiss mathematician Leonhard Euler.

So what did the ancient Greeks call the ratio? They probably did not have a handy abbreviation. Veltman explains:

> It appears that even the Greek mathematicians themselves did not use their letter pi to represent the circle ratio. Several ancient cultures did their math in sentence form with little or no abbreviation or symbolism. It is amazing how much they achieved without such an aid. Archimedes, born about 287 B.C., is said to have determined that the value of pi was between $3^{10}/_{71}$ and $3\frac{1}{7}$ (or in our decimal notation, between 3.14084 and 3.142858).

In the centuries that followed, mathematicians in other countries produced even more precise calculations: Around A.D. 150, Ptolemy of Alexandria weighed in with his value of 3.1416. Around A.D. 480, a Chinese man, Tsu Ch'ung-chih, improved the figure to 3.1415929, correct to the first six numbers after the decimal point.

Why is our mathematical vocabulary a seemingly random hodgepodge of Greek and Latin terminology? Why do we say the Latin-derived "circumference" (originally meaning "to run or move around") rather than the Greek "periphery"? Diane McCulloch, a mathematician at the Mount de Chantal Academy in Wheeling, West Virginia, explains:

> The Greeks were more avid mathematicians than the Romans, who preferred the practical uses and didn't have much time for the analytical aspect of mathematics. Thank goodness the writings of the Greeks were preserved by the Islamic scholars. We have access to the ancient Greek mathematical work because these Islamic scholars established libraries in Spain; when they were eventually driven out of Spain, their books were translated into Latin and then into other European languages, which themselves tended to be derived from Latin.
>
> Therefore, we use the word "triangle" when the Greeks would probably have used the word "trigon," as in "polygon," "octagon," etc.

Submitted by Dennis Kingsley of Goodrich, Michigan.

DO PENGUINS HAVE KNEES? 81

Why Can't You Buy Macadamia Nuts in Their Shells?

Macadamia nuts do have shells. But selling them in their shells would present a serious marketing problem. Only Superman could eat them. According to the Mauna Loa Macadamia Nut Corporation, the largest producer of macadamias in the world, "It takes 300-pounds-per-square-inch of pressure to break the shell."

After macadamias are harvested, the husks are removed, and then the nuts are dried and cured to reduce their moisture. The drying process helps separate the kernel from the shell; without this separation, it would be impossible to apply the pressure necessary to shatter the shell without pulverizing the contents. The nuts then pass through counter-rotating steel rollers spaced to break the shell without shattering the nutmeat.

Of course, one question remains. Why did Mother Nature bother creating macadamias when humans and animals (even raging rhinos) can't break open the shells to eat them without the aid of heavy machinery?

Submitted by Herbert Kraut of Forest Hills, New York.

If Heat Rises, Why Does Ice Form on the Top of Water in Lakes and Ponds?

Anyone who has ever filled an ice-cube tray with water knows that room temperature water decreases in density when it freezes. We also know that heat rises. And that the sun would hit the top of the water more directly than water at the bottom. All three scientific verities would seem to indicate that ice would form at the bottom, rather than the top, of lakes and ponds. "What gives?" demand *Imponderables* readers.

You may not know, however, what Neal P. Rowell, retired professor of physics at the University of South Alabama, told us:

DAVID FELDMAN

Water is most dense at 4 degrees Centigrade (or 39.2 degrees Fahrenheit). This turns out to be the key to the mystery of the rising ice. One of our favorite scientific researchers, Harold Blake, wrote a fine summary of what turns out to be a highly technical answer:

> As water cools, it gets more dense. It shrinks. It sinks to the bottom of the pond, lake, rain barrel, wheelbarrow, or dog's water dish. But at 4 degrees Centigrade, a few degrees above freezing, the water has reached its maximum density. It now starts to expand as it gets cooler. The water that is between 4 degrees Centigrade and zero Centigrade (the freezing point of water) now starts to rise to the surface. It is lighter, less dense.
>
> Now, more heat has to be lost from the water at freezing to form ice at freezing. This is called the "heat of fusion." During the freezing process, ice crystals form and expand to a larger volume, fusing together as they expand, and using more freezing water to "cement" themselves together. The ice crystals are very much lighter and remain on the surface.
>
> Once the surface is frozen over, heat dissipates from the edges and freezing is progressive from the edges. When the unfrozen core finally freezes, there is tremendous pressure exerted from the expansion, and the ice surface or container sides yield, a common annoyance with water pipes.

Once the top layer of the lake or pond freezes, the water below will rarely reach 0 degrees Centigrade; the ice acts effectively as insulation. By keeping the temperature of the water below the ice between 0 and 4 degrees Centigrade, the ice helps some aquatic life survive in the winter when a lake is frozen over.

The strangest element of this ice Imponderable is that since water at 4 degrees Centigrade is at its maximum density, it always expands when it changes temperature, whether it gets hotter or cooler.

Submitted by Richard T. Mitch of Dunlap, California. Thanks also to Kenneth D. MacDonald of Melrose, Massachusetts; R. Prickett of Stockton, California; Brian Steiner of Charlotte, North Carolina; and John Weisling of Grafton, Wisconsin.

What Happens to the 1,000 or More Prints After Films Have Finished Their Theatrical Runs?

Distribution strategies for films vary dramatically. A "critics' darling," especially a foreign film or a movie without big stars, might be given a few exclusive runs in media centers like New York City and Los Angeles. The film's distributor prays that word-of-mouth and good reviews will build business so that it can expand to more theaters in those cities and later be distributed throughout the country.

"High-concept" films, particularly comedies and action films whose plot lines can be easily communicated in short television commercials, and films starring "bankable" actors are likely to be given broad releases. By opening the film simultaneously across the country at 1,000 to 2,000 or more theaters, the film studios can amortize the horrendous cost of national advertising. But the cost of duplicating 2,000 prints, while dwarfed by marketing costs, is nevertheless a major expense.

A run-of-the-mill horror film from a major studio, for example, might open in 1,500 theaters simultaneously. The usual pattern for these films is to gross a considerable amount of money the first week and then fall off sharply. A horror film without good word-of-mouth might be gone from most theaters within four weeks. Of course, studios will release the film on videotape within a year, but what will they do with those 1,500 prints?

The first priority is to ship the prints overseas. Most American films are released overseas after the American theatrical run is over. Eventually, those American prints are returned to the United States.

And then they are destroyed and the silver is extracted from the film and sold to precious metals dealers. The studios have little use for 1,500 scratchy prints.

Mark Gill, vice-president of publicity at Columbia Pictures, told *Imponderables* that his company keeps twenty to thirty prints of all its current releases indefinitely. The film studios are

DAVID FELDMAN

aware of all of the movies from the early twentieth century that have been lost due to negligence—some have deteriorated in quality but others are missing simply because nobody bothered keeping a print. With all of the ancillary markets available, including videotape, laser disc, repertory theaters, cable television, and syndicated television, today's movies are unlikely to disappear altogether (though we can think of more than a few that we would like to disappear). But the problem of print deterioration continues.

Submitted by Ken Shafer of Traverse City, Michigan. Thanks also to John DuVall of Fort Pierce, Florida.

Why Is Balsa Wood Classified as a Hardwood When It Is Soft? What Is the Difference Between a Softwood and a Hardwood?

Call us naive. But we thought that maybe there was a slight chance that the main distinction between a softwood and a hardwood was that hardwood was harder than softwood. What fools we are.

Haven't you gotten the lesson yet? LIFE IS NOT FAIR. Our language makes no sense. The center will not hold. Burma Shave.

Anyway, it turns out that the distinction between the two lies in how their seeds are formed on the tree. Softwoods, such as pines, spruce, and fir, are examples of gymnosperms, plants that produce seeds without a covering. John A. Pitcher, director of the Hardwood Research Council, told *Imponderables* that if you pull one of the center scales back away from the stem of a fresh pine cone, you'll see a pair of seeds lying side by side. "They have no covering except the wooden cone."

Hardwoods are a type of angiosperm, a true flowering plant that bears seeds enclosed in capsules, fruits, or husks (e.g., olives, lilies, walnuts). Hardwoods also tend to lose their leaves in temperate climates, whereas softwoods are evergreens; but in tropical climates, many hardwoods retain their leaves.

While it is true that there is a tendency for softwoods to be softer in consistency (and easier to cut for commercial purposes), and for hardwoods to be more compact, and thus tougher and denser in texture, these rules of thumb are not reliable. Pitcher enclosed a booklet listing the specific gravities of the important commercial woods in the U.S. He indicates the irony:

> At 0.16 specific gravity, balsa is the lightest wood listed. At a specific gravity of 1.05, lignumvitae is the heaviest wood known. Both are hardwoods.

Do Earlobes Serve Any Particular or Discernible Function?

Our authorities answered as one: Yes, earlobes do serve a particular function. They are an ideal place to hang earrings.

Oh sure, there are theories. Ear, nose, and throat specialist Dr. Ben Jenkins of Kingsland, Georgia, remembers reading about a speculation that when our predecessors walked on four feet, our earlobes were larger "and that they fell in[ward] to protect the ear canal." Biologist John F. Hertner recounts another anthropological theory: that earlobes served as "an ornament of interest in sexual selection."

Doctors and biologists we confront with questions like these about seemingly unimportant anatomical features are quick to

DO PENGUINS HAVE KNEES? 87

shrug their shoulders. They are quite comfortable with the notion that not every organ in our body is essential to our well-being and not every obsolete feature of our anatomy is eliminated as soon as it becomes unnecessary.

Actually, the opposite is closer to the truth. Anatomical features of earlier humankind tend to stick around unless they are an obvious detriment. As Professor Hertner puts it,

> Nature tends to conserve genetic information unless there is selection pressure against a particular feature. Our bodies serve in some respect as museums of our evolutionary heritage.

Submitted by Dianne Love of Seaside Park, New Jersey.

Why Does Butter Get Darker and Harder in the Refrigerator After It Is Opened?

Butter discolors for the same reason that apples or bananas turn dark—oxidation. And although butter doesn't have a peel to protect it from the ravages of air, it does have a snug wrapper surrounding it until it is first used by the consumer. Only after the wrapper is eliminated or loosened does the butter darken.

Why does it get harder? The cold temperature in the refrigerator causes the moisture in the butter to evaporate. Many other foods, such as peanut butter and onion dip, become less plastic when refrigerated because of evaporation of liquid.

Submitted by Mitchell Hofing of New York, New York.

Why Are Dance Studios Usually Located on the Second Floor of Buildings?

Truth be told, the quality of Imponderables we receive on call-in talk shows is usually distinctly inferior to the stuff in our mailbag. But Los Angeles talk-show host Carol Hemingway has an imaginative gang of listeners, and this question was posed to us on her gabfest.

We loved it. This was an Imponderable deeply embedded in our subconscious. We have traveled through small towns with only one two-story commercial building in them. If there was a dance studio in town, it was located in that building. On the second floor.

We recently saw film critic Roger Ebert wax nostalgic about his first date. He asked his friend to a dance at Thelma Lee Ritter's studio in Urbana, Illinois. Unfortunately, the dance was canceled. But not all was lost—Ms. Ritter's second-floor studio was located above the Princess Theater. So Roger and his date went to the movies.

In New York, dance studios are invariably on the second floor as well. Not on the twenty-seventh floor. The second.

The best explanation we could figure out is that rents must obviously be cheaper on higher floors. Dance studios presumably have little walk-in, impulse business. Even folks with happy feet are unlikely, on the spur of the moment, to decide they instantly must have tango lessons. All of our sources indicated that these factors were crucial, but other considerations were important, too.

We heard from Connie Townsend, national secretary of the United States Amateur Ballroom Dancers Association in Baltimore, Maryland, who "reviewed the many studios with which I am familiar and was somewhat surprised to find that, indeed, most are located on second floors." Townsend notes that all the exceptions she could think of were built specifically as dance studios by their owners.

Frank Kiley, a former ballroom dance instructor and former licensee of 1,800 ballrooms nationwide, provides a historical perspective and an architectural answer:

> Previous to 1980, most studio-type ballrooms had to have twelve-foot ceilings to position loudspeakers for maximum effect. Some studios had fourteen- or eighteen-foot-high ceilings for best audio results . . .
>
> Most second-floor buildings in major cities had to use special curtains to subdivide ballroom classes and higher floors tended to have larger windows and smaller pillars in their structural design.

Kiley's reasons were echoed by Vickie Sheer, executive director of the Dance Educators of America, but she added several others as well:

> When I taught for thirty-seven years, my studios were always located on the second floor. My reason was to deter anyone from coming in from street level. If people climbed a flight of stairs, there must be [genuine] interest. Also, the second-floor placement kept out annoying children opening and closing doors and being pests.

DAVID FELDMAN

In winter, the heat in a building goes up and the cold air does not blow in, as on street level. Usually there is more square footage on a second-floor than a street-level . . .

Kiley, still a major copyright owner in the ballroom industry, notes that dance studios are invading shopping malls, where many have landed on the ground floor.

Submitted by a caller on the Carol Hemingway show, KGIL-AM, Los Angeles, California.

Why Are 25-Watt Light Bulbs More Expensive Than 40-, 60-, 75-, and 100-Watt Bulbs?

The old rule of supply and demand takes effect here. You don't always get what you pay for. Richard H. Dowhan, manager of public affairs for GTE, explains:

> The higher-wattage, 40-, 60-, 75-, and 100-watt light bulbs are manufactured in huge quantity because they are in demand by consumers. The 25-watt light bulb has limited uses, therefore fewer bulbs are manufactured and you don't get the inherent cost advantage of large productions runs.
>
> Secondly, in order to make it worthwhile for the retailer to stock a slow mover, which takes up shelf and storage space for longer periods of time, you increase the profit margin. These two factors result in a higher price.

Submitted by Alan Snyder of Palo Alto, California.

Why Are Water Towers Built So High?

We have passed through small towns and cities where the water tower is by far the highest structure in sight. The name of the city is often emblazoned around the surface of the mighty edifice.

But why are the water towers necessary anyway when most communities have reservoirs? And why are they so tall?

We got our answer from Dr. Paul J. Godfrey, director of the Water Resources Center at the University of Massachusetts at Amherst:

> The task of providing water through a municipal distribution system requires both sufficient volume to meet normal consumption, emergency consumption (such as for fighting major fires), and sufficient pressure to operate household and industrial devices.
>
> The water tower provides, by its volume, a reservoir that can meet short-term needs during periods of high water use, usually in the morning and around dinner time, and allows the various sources of supply, reservoirs, or wells, to catch up during lower demand. The volume chosen for a facility usually assumes that a sufficient volume must be available to meet simultaneous demand during the high consumption period and a major fire. All of these functions do not require a particular water height.
>
> But water pressure is provided by the height of the tower. Water seeks its own level. For example, if we fill a water hose with water and hold each end of the hose at exactly the same level, no water flows out. But if one end of the hose is raised, water flows out of the other end. . . . The more water, and hence weight, above the lower end, the greater the force of water flowing out of the hose. The role of the water tower is to provide an elevated weight of water sufficient to provide adequate pressure at all outlets in the system.

The alternative would be to install electrical pumps to force water out of other reservoirs, but this is an inefficient technology. As Peter Black, president of the American Water Resources, told *Imponderables*, with a pump system "You would have to activate the pump every time anybody wanted any water."

The water tower may be lower tech than an electrical pump, but with the precautions outlined by Dr. Godfrey, it does the job well over any terrain:

DAVID FELDMAN

To create adequate pressure for all parts of a municipality, the water tower must be higher than all the municipality's water taps, sufficiently high to create fire-fighting pressure at all hydrants. In some cities, the municipal supply does not provide enough pressure for tall buildings, so booster pumps and another storage tank on top of the building supplements pressure. [In New York and other big cities, standpipes are placed in front of tall buildings to insure that water can be delivered to the top of the building.]

Water pressure in all areas of the municipality must be carefully controlled. Low pressure will be dangerous in a serious fire and will produce complaints from those who like brisk showers. High pressure will wear out valves and gaskets faster and cause excessive system leakage. Towns with high hills will often have squat water tanks on the highest hill and install pressure regulators to reduce pressure in the valleys. Towns with no hills must compensate by building an artificial hill, the water tower, which is higher than the tallest building.

So those water towers aren't so high just to serve as a monument to the ego of the mayor. The tall water tower ends up saving energy and money. But as Jay H. Lehr, executive director of the Association of Ground Water, explains, even our hero, the water tower, isn't perfect: "Of course, there is no free lunch, as electrical energy is used in pumping the water into the storage tower in the first place."

Submitted by Cheri Klimes of Cedar Rapids, Iowa. Thanks also to George Armbruster of Greenbelt, Maryland; Gary Moore of Denton, Texas; and Jason and Bobby Nystrom of Destrehan, Louisiana.

Why Are Some Watermelon Seeds White and Some Black?

Most of you probably think all watermelons contain black and white seeds. It's time for some serious consciousness raising. And Gary W. Elmstrom, professor of horticulture at the Institute of Food and Agricultural Sciences at the University of Florida, is just the man to do it:

> Different varieties of watermelons have an array of different-colored seeds. Color of mature seed can vary from almost white to black depending upon variety. A watermelon variety named "Congo" has white seeds and "Jubilee" has black seeds. There are also genes for red-colored seeds in watermelons such as in a variety called red-seeded citron. . . . Seed color in watermelons is genetically determined, just as eye color is in humans.

So is one melon bearing both white and black seeds the equivalent of a human with one blue eye and one brown eye? Not quite. Professor Elmstrom explains: "Just as blue-eyed babies turn into brown-eyed children, so do white seeds, barring pollination or fertility problems, end up as black ones."

Submitted by John K. Aldrich of Littleton, Colorado. Thanks also to J. R. Shepard of Plantation, Florida.

What Is the Purpose of the Holes Near the End of Electric Plug Prongs?

Most of our hardware sources knew the answer to this Imponderable, which, judging by our mail, is high in the consciousness of the *spiritus mundi*. Ed Juge, director of market planning for Radio Shack, provided a succinct answer:

DAVID FELDMAN

The holes near the ends of AC plug prongs are there to mate with spring-loaded pins found in some of the better wall sockets, to help make a good connection, and to keep the plug from falling out of the socket.

Submitted by Venia Stanley of Albuquerque, New Mexico. Thanks also to William C. Stone of Dallas, Texas; George A. Springer of San Jose, California; Barry Cohen of Thousand Oaks, California; Jesse D. Maxenchs of Sunnyvale, California; and Rory Sellers of Carmel, California.

Why Do Paper Mills Smell So Bad?

Rose Marie Kenny, of Hammermill Papers, properly chided us for posing the wrong Imponderable. Paper mills *don't* generate offensive odors, she reminded us. Pulp mills do.

You're right, Ms. Kenny. Excuse us if we were too busy holding our noses to notice that rotten eggs were emanating from a pulp mill, not a producer of finished paper products.

We asked Stephen Smulski, a professor of wood science and technology at the University of Massachusetts at Amherst, to explain how lovely-scented trees turn into foul-smelling pulp:

> Paper is a thin sheet of tangled wood fibers, each of which is like a long, hollow straw with tapering, closed ends, and about ½ inch in length. In the standing tree, wood fibers, which consist mainly of cellulose, are held together by and embedded in an adhesive-like material called lignin.
>
> In order to isolate the individual wood fibers needed for making paper, chips of solid wood are treated with chemicals that

DAVID FELDMAN

selectively dissolve only lignin in a process called pulping. Of the several wood pulping processes, only the kraft sulfate process emits the rotten cabbage smell associated with pulp mills.

In this process, sodium hydroxide (NaOH) and sodium sulfide (Na$_2$S) are used to dissolve lignin. Though these chemicals are recovered and reused in a closed cycle, and the gases vented from the process scrubbed clean with state-of-the-art pollution control technology, tiny amounts of sulfur still escape into the air.

Unfortunately for all of us, not least the citizens of the towns in which the pulp mill is located, you don't need to be a dog to sniff out the scent of sulfur compounds. Doug Matyka, a public relations manager at Georgia-Pacific's Tissue, Pulp and Bleached Board division, told *Imponderables* these chemicals are "readily noticeable even at levels far below one part per million."

But if a paper company decides to locate a plant in your town, don't despair before you ferret out a few facts. Not all paper plants make their own pulp; many buy their pulp from a pulpmaking facility or from a free-standing mill that makes only pulp.

Also, many papermaking methods don't require these sulfur compounds. The olfactory culprit is the kraft-pulping process ("kraft" derives from the German term meaning "strong"). Kraft paper is best known for making supermarket shopping bags and corrugated cardboard boxes. If the other pulping processes were capable of producing the strength performance of kraft paper at a decent price, don't you think that companies would employ them? After all, even executives of paper companies have to smell the stuff, too.

Lest we seem to be picking on the pulp industry, a few fun facts from our paper sources will help you understand their dilemma:

1. You can't make paper without pulp. According to Kenny, 80 percent of a sheet of paper is made out of pulp.

2. High-tech scrubbers have diminished the odor problem considerably. And more than many other industries, pulp mills conform to EPA air and water pollution standards. The smell of sulfur compounds is more offensive than dangerous.

3. Many other things besides pulp mills produce the smell. Doug Matyka notes that exhaust from vehicles with catalytic converters sometimes smells like mini-pulp mills. And the same smell occurs during natural organic decay. After all, Doug reminds us, the original descriptive phrase "you smell like a rotten egg" comes not from pulp mills but rotten eggs.

Yeah, sure, but it's more fun to pick on heavy industry than a chicken.

Submitted by Barry Long of Alexandria, Virginia.

How Are Lane Reflectors Fastened onto the Road So That They Aren't Moved or Crushed?

Most of us, on occasion, have accidentally run over a lane reflector. The little bump is always upsetting. Have we moved the reflector? Have we hurt the reflector? Have we hurt our tires?

Don't worry too much about the reflectors, for you are unlikely to dislodge them. A two-part epoxy cement is used to fasten them to the road surface.

But the nasty little secret of the reflectors' durability is that they are recessed into the road surface to prevent movement and designed in the shape of a two-sided ramp to avoid getting crushed. George E. Jones, highway engineer at the National Highway Institute, told *Imponderables* that

> if you will look closely you will notice a groove about a foot long cut at a downward sloping angle in the pavement. The reflector is then cemented flush with the pavement surface.

Well, not quite flush, and for a good reason. Amy Steiner of the American Association of State Highway and Transportation Officials explains:

... they do protrude slightly so that they can catch the light emitted from the headlights. Because of this protrusion, they do sustain damage from vehicles driving over them (particularly from snowplow blades). Reflectors require a fair amount of maintenance.

As for your tires, our authorities agree they are safe even if you hit twenty reflectors on the center line of the road. Just make sure you stay in your lane, or more than your tires are in jeopardy.

Submitted by Eric Hartman of Spring Grove, Pennsylvania.

Why Does Nabisco Put the Tiny Picture of Niagara Falls on Its Shredded Wheat Box?

At the beginning of the twentieth century, shredded wheat biscuits were produced at The Palace of Light, a ten-acre site bordering Niagara Falls, New York. According to Michael Falkowitz, director of industry and trade relations for Nabisco Brands, The Palace of Light itself became a tourist mecca and served as the marketing image for the Shredded Wheat Company:

> The plant was decked out in marble tile and glass, and was air conditioned. It was visited by more than one hundred thousand wide-eyed tourists every year. The great falls gave meaning to what was billed as "the cleanliest" product—Shredded Wheat (and Triscuit)—produced in the cleanliest food factory.

DAVID FELDMAN

When the Nabisco Biscuit Company acquired the Shredded Wheat Company in 1928, it didn't mess with packaging that was working just fine. It did mess with The Palace of Light, though. No longer state-of-the-art, the Niagara Falls plant was abandoned by Nabisco before World War II.

Submitted by Sister Anne Joan of Boston, Massachusetts.

Why Do Automobile Batteries Have to Be So Heavy? Why Can't They Be Miniaturized?

Of course, most consumers would prefer car batteries to be AA-size. If a car stalled, a driver could just reach into the glove compartment and pull out a little battery that had been re-charged at home.

Automobile manufacturers also want to downsize batteries. Any heavy material, whether it is the steel in the body of a car or the engine and cylinders, interferes with achieving better gasoline mileage.

Battery manufacturers have responded. In some cases batteries are half the size they were twenty years ago. But alas, don't look forward to AA-sized car batteries in the foreseeable future. As Stephen Bomer of the Automotive Battery Charger Manufacturers wrote to *Imponderables*, high-density lead plates are a major component of a battery: "No substitute for lead has been found that can do the job or generate the voltage required."

H. Dale Millay, a staff research engineer for Shell Oil, told *Imponderables* that the greater the surface area of lead in the battery, the easier it is to generate power. Millay claims that we have already paid the price for downsizing batteries: Although modern batteries are good at cold starts, they have low reserve capacities. Translation: They don't last as long as they might under strain.

We received our most emphatic endorsement of the heavy battery from John J. Surrette, vice-president of Rolls Battery Engineering:

> The thinner you make the plates in a battery, the lesser the material inside. . . . The heavier the material, the more rugged the batteries are and the longer they will last. When you use thinner plates . . . this lessens the amount of ampere hour capacity. When heavier material is used, like we do in marine and industrial applications, it results in considerably longer life and less exposure [to the elements], which reduces the chance of plates buckling in hard service or the active material shedding from the positive grids.
>
> . . . Miniaturized batteries would probably be preferable but would stand little or no abuse or neglect.

Rolls's marketing strategy is to emphasize the *heaviness* of its battery. It boasts a marine battery with 1/8"-thick positive plates (in contrast, some car batteries have plates as thin as .055 inches, which Surrette believes is too fragile to withstand abuse or neglect).

So, Becky, the car battery is one case where you *don't* want to get the lead out.

Submitted by Becky Brown of Iowa City, Iowa.

DAVID FELDMAN

Why Do Automatic Icemakers in Home Freezers Make Crescent-Shaped Pieces Rather Than Cubes?

We cannot be dispassionate about this subject. We hate crescent-shaped cubes. They are so long they get stuck in iced-tea glasses, making it impossible to load enough ice to cool the drink sufficiently. Even in wider glasses, they are too ungainly to stack. And they are too big to pop into one's mouth comfortably.

We contacted the company that pioneered the automatic icemaker for home freezers, the Whirlpool Corporation, and prepared for a battle. What excuse would it trot out to justify banishing the ice cube to oblivion?

Much to our surprise, the manager of Whirlpool's appliance information service, Carolyn Verweyst, disarmed us with her compassion and empathy. She, too, dislikes the crescent-shaped cubes, if only because they make her job harder: She says that Whirlpool gets more complaints about the shape of the ice than anything else about their refrigerator-freezers.

Verweyst and our other appliance sources confirm that there is only one reason why these cubes are crescent-shaped: Any other shape tends to stick to the mold instead of releasing to the ice bin. When first developing the automatic icemaker, Whirlpool experimented with many different shapes but found that any ice with straight edges simply would not release properly.

In fact, Verweyst says that Whirlpool gave an outside think tank a project to come up with a perfect shape for ice molded by an automatic icemaker. Its conclusion: The best shape was a crescent.

Submitted by Emily Sanders and Elizabeth Gaines of Montgomery, Alabama.

What Are Those Computer Scrawls (Similar to Universal Product Codes) Found on the Bottom Right of Envelopes? How Do They Work?

You'd better get used to those scrawls. If the United States Postal Service reaches its goal, every single letter and package sent through them will have bar codes by the year 1995.

The series of vertical lines on the lower-right portion of first-class envelopes is meant for the "eyes" of OCRs, high-speed optical character readers. The Postal Service calls this specific bar code configuration POSTNET, short for Postal Numeric Encoding Technique.

OCRs are now capable of "reading" typewritten or hand-printed addresses and spraying bar codes on an envelope. The bar code readers are considerably less sophisticated and less expensive machines than OCRs (or for that matter, much less expensive than hiring humans over the long haul). By automating the sorting process, the postal service speeds mail delivery and saves more than a buck or two at the same time.

Can a human being interpret the code sprayed by the OCRs? Absolutely, although it's a little tricky. If you look carefully, you will see that all the bars are two heights—either full bars or half-sized. The tall bars represent (binary) ones; the short bars represent (binary) zeros. The bars on the far left and far right, always full bars, are not part of the code, and are there merely to frame the other numbers.

DAVID FELDMAN

All the rest of the bars and half-bars are arranged in groups of five. Each group of five bars represents one of the ZIP code digits, and all numbers are always expressed by two full bars and three half-bars. You can figure out which number the bars represent by noting which of the five positions contain full bars. Here is the code (remember, one equals a full bar; zero equals a half bar):

$$11000 = 0 \quad 01010 = 5$$
$$00011 = 1 \quad 01100 = 6$$
$$00101 = 2 \quad 10001 = 7$$
$$00110 = 3 \quad 10010 = 8$$
$$01001 = 4 \quad 10100 = 9$$

These ten combinations can express all possible ZIP codes.

The bar code readers, working from left to right, add the values of the two full bars for each group of five to arrive at the proper ZIP code. Each of the five bar codes always has the same numerical assignment. From left to right, they are 7, 4, 2, 1, and 0. So if one of the digits of the ZIP code is 7, it will be expressed by making the first bar (7) and fifth (0) bar code full. If you add 7 and 0, you get seven (we hope). To express the number 6, the second (4) and third (2) bars in the group would be full height.

Got it? Good, because we have one more feature to confuse you with. After the nine- or five-digit code is sprayed, one other group of five bars is added to the right of the last ZIP code digit —the "correction character." To arrive at this number, add all the digits in the ZIP code. The *Imponderables* post office box, for example, is in the 90024 ZIP area. By adding all the individual digits in the ZIP code, we arrive at the sum of fifteen. To calculate the proper correction character, the bar code reader subtracts this sum from the next highest multiple of ten—in this case, 20. (If the sum were 38, it would be subtracted from 40.) The remainder, five, is expressed as any other five would be in the POSTNET system—with the second and fourth bar being full height.

What is truly remarkable about the POSTNET system is how fast the OCRs can operate, and how they are capable of converting five-digit ZIP codes into the nine-digit ones that the postal system prefers. Reader Harold E. Blake, an expert on OCRs whose expertise was invaluable in writing this entry, summarizes how our letter was processed and sent along its merry way to him in the mellifluously named town of Zephyrhills, Florida:

> This letter went through an OCR at the rate of nine per second. In about one-ninth of a second, the face of the letter was read and reassembled in a computer register. A multimegabyte memory was searched (sort of like an electronic ZIP code directory for the United States) and my post office box number was verified to be identical to the four in the ZIP + 4 (this is not always the case with box numbers). The 33539 was matched with Zephyrhills as a valid association. Then, a signal went from the computer to an A.B. Dick bar code printer, and "spriff," several hundred ink dots got sprayed on this letter as it moved along faster than the eye could track it.
>
> All in 110 milliseconds.

POSTNET codes are now commonly preprinted on business reply and courtesy reply envelopes by mass mailers so that the mail will bypass OCRs altogether and go directly to a bar code sorter. The chance of a misdirected letter is greatly reduced. It might not matter anymore if you "accidentally" put the insert of your utility bill upside-down so that no print shows through the address window. That bar code emblazoned on the return envelope is telling a bar code sorter the nine-digit ZIP code of the utility; chances are, the bill will arrive at its intended destination. Ain't progress grand?

Submitted by George Persico of Thiells, New York. Thanks also to Tom Emig of St. Charles, Missouri; Cynthia J. Gould of Fairhaven, Massachusetts; Harold Fair of Bellwood, Illinois; Bob Peterson, United States Air Force, APO New York; William J. Feole, Alameda, California; Millicent Brinkman of Thornwood, New York; Kristina Castillo of Williamsville, New York; Herman London of Poughkeepsie, New York; Debbie DiAntonio of Malvern, Pennsylvania; and many others.

DAVID FELDMAN

If Water Is Composed of Two Parts Hydrogen and One Part Oxygen, Both Common Elements, Why Can't Droughts Be Eliminated by Combining the Two to Produce Water?

We could produce water by combining oxygen and hydrogen, but at quite a cost financially and, in some cases, environmentally.

Brian Bigley, senior chemist for Systech Environmental Corporation, says that most methods for creating water are impractical merely because "you would need massive amounts of hydrogen and oxygen to produce even a small quantity of water, and amassing each would be expensive." Add to this the cost, of course, the labor and equipment necessary to run a "water plant."

Bigley suggests another possible alternative would be to obtain water as a byproduct of burning methane in an oxygen atmosphere:

Again, it's a terrible waste of energy. Methane is a wonderful fuel, and is better used as such, rather than using our supply to produce H_2O. It would be like giving dollar bills to people for a penny to be used as facial tissue.

The most likely long-term solution to droughts is desalinization. We already have the technology to turn ocean water into drinking water, but it is too expensive now to be commercially feasible. Only when we see water as a valuable and limited natural resource, like oil or gold, are we likely to press on with large-scale desalinization plants. In northern Africa, water for crops, animals, and drinking is not taken for granted.

Submitted by Bill Irvin III of Fremont, California.

Why Does Your Voice Sound Higher and Funny When You Ingest Helium?

The kiddie equivalent of the drunken partygoer putting a lampshade on his head is ingesting helium and speaking like a chipmunk with a caffeine problem. When we saw *L.A. Law*'s stolid Michael Kuzak playing this prank, we were supposed to be smitten with his puckish, fun-loving, childlike side. We were not convinced.

Still, many *Imponderables* readers want to know the answer to this question, so we contacted several chemists and physicists. They replied with unanimity. Perhaps the most complete explanation came from George B. Kauffman:

> Sound is the sensation produced by stimulation of the organs of hearing by vibrations transmitted through the air or other mediums. Low-frequency sound is heard as low pitch and higher frequencies as correspondingly higher pitch. The frequency (pitch) of sound depends on the density of the medium through which the vibrations are transmitted; the less dense the medium,

the greater the rate (frequency) of vibration, and hence, the higher the pitch of the sound.

The densities of gases are directly proportional to their molecular weights. Because the density of helium (mol. wt. 4) is much less than that of air, a mixture of about 78 percent nitrogen (mol. wt. 28) and about 20 percent oxygen (mol. wt. 32), the vocal cords vibrate much faster (at a higher frequency) in helium than in air, and therefore the voice is perceived as having a higher pitch.

The effect is more readily perceived with male voices, which have a lower pitch than female voices. The pitch of the voice [can] be lowered by inhaling a member of the noble (inert) gas family (to which helium belongs) that is heavier than air, such as xenon (mol. wt. 131.29). . . .

Brian Bigley, a chemist at Systech Environmental Corporation, told *Imponderables* that helium mixtures are used to treat asthma and other types of respiratory ailments. Patients with breathing problems can process a helium mixture more easily than normal air, and the muscles of the lungs don't have to work as hard as they do to inhale the same volume of oxygen.

Submitted by Jim Albert of Cary, North Carolina. Thanks also to James Wheaton of Plattsburg AFB, New York; Nancy Sampson of West Milford, New Jersey; Karen Riddick of Dresden, Tennessee; Loren A. Larson of Altamonte Springs, Florida; and Teresa Bankhead of Culpepper, Virginia.

Why Is the French Horn Designed for Left-Handers?

We hope that this Imponderable wasn't submitted by two left-handers who learned the instrument because they were inspired by the idea that an instrument was finally designed specifically for them. If so, Messrs. Corcoran and Zitzman are in for a rude awakening.

If we have learned anything in our years toiling in the mine-fields of Imponderability, it is that *nothing* is designed for left-handers except products created exclusively for lefties that cost twice as much as right- (in both senses of the word) handed products.

In case the premise of the Imponderable is confusing, the French horn is the brass wind instrument with a coiled tube—it looks a little like a brass circle with plumbing in the middle and a flaring bell connected to it. The player sticks his or her right hand into the bell itself and hits the three valves with the left hand. So the question before the house is: Why isn't the process reversed, with the difficult fingering done by the right hand?

You've probably figured it out already. The original instrument had no valves. Dr. Kristin Thelander, professor of music at the University of Iowa School of Music and a member of the International Horn Society, elaborates:

> In the period 1750–1840, horns had no valves, so the playing technique was entirely different from our modern technique. The instruments were built with interchangeable crooks which placed the horn in the appropriate key for the music being played, and pitches lying outside of the natural harmonic series were obtained by varying degrees of hand stopping in the bell of the horn.
>
> It was the right hand which did this manipulation in the bell of the horn, probably because the majority of people are right-handed [another theory is that earlier hunting horns were designed to be blown while on horseback. The rider would hold the instrument with the left hand and hold the reins with the right hand].
>
> Even when the valves were added to the instrument, a lot of

DAVID FELDMAN

hand technique was still used, so the valves were added to the left-hand side.

On the modern French horn, this hand technique is no longer necessary. But so many generations grew up with the old configuration that the hand position remains the same. Inertia triumphs again, even though it would probably make sense for right-handers to use their right hands on the valves. But fair is fair: Lefties have had to contend with all the rest of the right-dominant instruments for centuries.

Most of our sources took us to task for referring to the instrument as the "French horn." In a rare case of our language actually getting simpler, the members of the International Horn Society voted in 1971 to change the name of the instrument from the "French Horn" to the "horn."

Why? Because the creators of the instrument never referred to it as the "French horn," any more than French diners order "French" dressing on their salads or "French" fries with their steak. As we mentioned earlier, the horn was the direct descendant of the hunting horn, which was very popular in France during the sixteenth and seventeenth centuries. The English, the same folks who screwed us up with the *cor anglais* or English horn started referring to the instrument as the "French horn" as early as the late seventeenth century, and the name stuck. Americans, proper lemmings, followed the English misnomer.

Submitted by Edward Corcoran of South Windsor, Connecticut.
Thanks also to Manfred S. Zitzman of Wyomissing,
Pennsylvania.

Why Do Milk Cartons Indicate "Open Other End" on One Side of the Spout and "To Open" on the Other When Both Sides Look Identical?

In our first book, *Imponderables,* we discussed why milk cartons are so difficult to open and close. To make a long story short, the answer: The current milk carton is extremely cheap to manufacture, and customers don't complain about the problem enough to motivate milk suppliers to change the packaging.

But three readers have written recently to ask about another milk carton conundrum that has always perplexed us. The top of the milk carton looks so symmetrical, it hardly *seems* to matter where you form the spout.

Alas, it does matter. Milk companies buy the paperboard for milk cartons unformed. Machines at the milk distributor form the paperboard into the familiar carton shape, seal the bottoms, fill the cartons with milk, and then seal the top. Bruce V. Snow, recently retired from the Dairylea Cooperative, explains:

> The machine is adjusted so that only one side of the gable (the "open this side" end) is sealed; when you pull the gable sides, the spout is exposed and opens. If you pull back the gable sides on the other end of the top, then squeeze the sides, nothing happens. The gable on that side stays sealed.

Why does it stay sealed? The secret, according to Dellwood dairy's Barbara Begany, is an ingredient called abhesive, "applied to the 'pour spout,' which makes it easier to open. Abhesive also prevents solid bonding of paper to paper as occurs on the 'open other end' side."

Submitted by Grayce Sine of Chico, California. Thanks also to Alice Conway of Highwood, Illinois, and Jeffrey Chavez of Torrance, California.

DAVID FELDMAN

Why Do We Feel Warm or Hot When We Blush?

We blush—usually due to an emotional response such as embarrassment (we, for example, often blush after reading a passage from our books)—because the blood vessels in the skin have dilated. More blood flows to the surface of the body, where the affected areas turn red.

We tend to associate blushing with the face, but blood is sent to the neck and upper torso as well. According to John Hertner, professor of biology at Nebraska's Kearney State College,

> This increased flow carries body core heat to the surface, where it is perceived by the nerve receptors. In reality, though, the warmth is perceived by the brain in response to the information supplied by the receptors located in the skin.

DO PENGUINS HAVE KNEES?

Because of the link between the receptors and the brain, we feel warmth precisely where our skin turns red.

Submitted by Steve Tilki of Derby, Connecticut.

During a Hernia Exam, Why Does the Physician Say, "Turn Your Head and Cough"? Why is the Cough Necessary? Is the Head Turn Necessary?

Although a doctor may ask you to cough when listening to your lungs, the dreaded "Turn your head and cough" is heard when the physician is checking for hernias, weaknesses or gaps in the structure of what should be a firm body wall.

According to Dr. Frank Davidoff, of the American College of Physicians, these gaps are most frequently found in the inguinal area in men, "the area where the tube (duct) that connects each testicle to the structures inside the body passes through the body wall." Some men are born with fairly large gaps to begin with. The danger, Davidoff says, is that

> Repeated increases in the pressure inside the abdomen, as from repeated and chronic coughing, lifting heavy weights, etc., can push abdominal contents into the gap, stretching a slightly enlarged opening into an even bigger one, and leading ultimately to a permanent bulge in contents out through the hernia opening.
>
> Inguinal hernias are obvious and can be disfiguring when they are large and contain a sizable amount of abdominal contents, such as pads of fat or loops of intestine. However, hernias are actually more dangerous when they are small, because a loop of bowel is likely to get pinched, hence obstructed, if caught in a small hernia opening, while a large hernia opening tends to allow a loop of bowel to slide freely in and out of the hernia "sac" without getting caught or twisted.
>
> Doctors are therefore particularly concerned about detecting inguinal hernias when they are small, exactly the situation in

DAVID FELDMAN

which they have not been obvious to the patient. A small inguinal hernia may not bulge at all when the pressure inside the abdomen is normal. Most small hernias would go undetected unless the patient increased the pressure inside the abdomen, thus causing the hernia sac to bulge outward, where it can be felt by the doctor's examining finger pushed up into the scrotum.

And the fastest, simplest way for the patient to increase the intra-abdominal pressure is to cough, since coughing pushes up the diaphragm, squeezes the lungs, and forces air out past the vocal cords. By forcing all the abdominal muscles to contract together, coughing creates the necessary increase in pressure.

If the physician can't feel a bulge beneath the examining finger during the cough, he or she assumes the patient is hernia-free.

And why do you have to turn your head when coughing? Dr. E. Wilson Griffin III, a family physician at the Jonesville Family Medical Center, in Jonesville, North Carolina, provided the most concise answer: "So that the patient doesn't cough his yucky germs all over the doctor."

Submitted by Jeffrey Chavez of Torrance, California. Thanks also to J. S. Hubar of Pittsburgh, Pennsylvania.

Why Is the "R" Trademark Symbol on Pepsi Labels Placed After the Second "P" in "Pepsi" Rather Than After the Words "Pepsi" or "Cola"?

We spoke to Chris Jones, Pepsi's charming manager of public affairs, who is by now used to our less than earth-shatteringly important questions. She told us that, legally speaking, the company could have put the registered mark wherever it wanted to.

But the company wanted to place the mark in the "Pepsi" rather than the "Cola." It seems that Pepsi has a competitor in the cola wars—its name escapes us at the moment—so they wanted to draw attention to the "P-word" rather than the "C-word."

DO PENGUINS HAVE KNEES?   115

Jones says that graphic designers felt that the mark after the second "p" looked better than placing it after the "i" in Pepsi: It made the design more symmetrical and didn't butt up against the hyphen after Pepsi.

Submitted by Tom Cunnifer of Greeley, Colorado.

Why Do Magazine and Newspaper Editors Force You to Skip Pages to Continue an Article at the Back of the Magazine/Newspaper?

We answered the question of why page numbers are missing from magazines in *Why Do Dogs Have Wet Noses?*. Now, from our correspondent Karin Norris: "It has always annoyed me to have to hold my place and search for the remainder of the article, hoping the page numbers will be there."

We hear you, Karin. In fact, one of the great pleasures of reading *The New Yorker* is the certainty that there will be no such jumps. We had always assumed that the purpose of jumps was to force you to go to the back of the book, thus making advertisements in nonprime areas of the paper or magazine more appealing to potential clients. Chats with publishers in both the newspaper and magazine field have convinced us that other factors are more important.

A newspaper's front page is crucial to newsstand sales. Editors want readers to feel that if they scan the front page, they can get a sense of the truly important stories of the day. If there were no jumps in newspapers, articles would have to be radically shortened or else the number of stories on the front page would have to be drastically curtailed.

Less obviously, magazine editors want what Robert E. Kenyon, Jr., executive director of the American Society of Magazine Editors, calls "a well-defined central section." Let's face it. Most magazines and newspapers are filled with ads, but with the possible exception of fashion and hobbyist magazines, readers are usually far more interested in articles. Magazine editors want to concentrate their top editorial features in one section to give at least the impression that the magazine exists as a vehicle for information rather than advertising. J. J. Hanson, chairman and CEO of The Hanson Publishing Group, argues that sometimes jumps are necessary:

> An article that the editor feels is too long to position entirely in a prime location will jump to the back of the book, thus permitting the editor to insert another important feature within the main feature or news "well." Many publishers try very hard to avoid jumps.
>
> The unhappiest version of a jump is one where an article jumps more than once so that instead of completing the article after the first jump, the reader reads on for a while and then has to jump again. That's almost unforgivable.

Hanson adds that another common reason for jumps in magazines, as opposed to newspapers, is color imposition:

> Most magazines do not run four-color or even two-color throughout the entire issue. Often the editor wants to position the major art treatment of his features or news items within that four-color section. In order to get as many articles as possible in that section, the editor sometimes chooses to jump the remaining portions of the story to a black and white signature.

Of course, advertising does play more than a little role in the creation of jumps. Most publications will sell clients just about any size ad they want. If an advertiser wants an odd-sized ad, one that can't be combined with other ads to create a full page of ads, editorial content is needed. It is much easier to fill these holes with the back end of jumps than to create special features to fill space. *The New Yorker* plugs these gaps with illustrations and funny clippings sent in by readers, which, truth be told, may be read more assiduously than their five-part book-length treatments on the history of beets.

Submitted by Karin Norris of Salinas, California.

What Does the EXEMPT Sign Next to Some Railroad Crossing Signs Mean?

EXEMPT signs are not intended for drivers of private cars, but rather for drivers of passengers for hire, school buses carrying children, or vehicles carrying flammable or dangerous materials. Ordinarily, these vehicles must stop not more than fifty feet or less than fifteen feet from the tracks of a railroad crossing, and their drivers are supposed to listen for signs of an approaching train, look in each direction along the tracks, and then proceed only if it is apparent no train is near.

But an exception to this federal regulation is granted to

> an industrial or spur line railroad grade crossing marked with a sign reading "EXEMPT." Such EXEMPT signs shall be erected only by or with the consent of the appropriate state or local authority.

According to the signage bible, the *Manual on Uniform Traffic Control Devices,* an EXEMPT sign informs the relevant

DAVID FELDMAN

drivers that "a stop is not required at certain designated grade crossings, except when a train, locomotive, or other railroad equipment is approaching or occupying the crossing or the driver's view of the sign is blocked."

Robert L. Krick, of the Federal Railroad Administration, told *Imponderables* that some states do not permit the use of EXEMPT signs or may attach additional meanings to them. And Krick makes it clear that an EXEMPT sign does not relieve any driver from the responsibility of determining that no train is approaching before entering a crossing. Krick emphasizes the motto of the FRA's Operation Lifesaver: "Trains Can't Stop; You Can."

Submitted by Tisha Land of South Portland, Maine.

In *Why Do Clocks Run Clockwise?*, we answered a myriad of questions about M&Ms. What food product seems to mystify North America today? Bubble gum, evidently.

What Flavor Is Bubble Gum Supposed to Be?

"No particular flavor," said a representative from Bubble Yum about its "regular" flavor.

"Fruit flavor—sort of a tutti frutti," responded an executive from Topps.

We hadn't encountered so much secrecy about ingredients since we pried the identity of the fruit flavors in Juicy Fruit gum from the recalcitrant folks at Wrigley.

In *When Do Fish Sleep?*, we discussed how bubble gum was invented by Walter Diemer, a cost accountant with the Fleer Corporation. Bruce C. Wittmaier, a relative of Mr. Diemer's, was the only source who would reply to our bubble gum question. And luckily, Wittmaier obtained his information directly from

Mr. Diemer. The main flavors in the original bubble gum: wintergreen, vanilla, and cassia.

Submitted by John Geesy of Phoenix, Arizona.

What Makes Bubble Gum Blow Better Bubbles Than Regular Chewing Gum?

All chewing gums consist of gum base, some form of sugar (or sorbitol in sugarless gums), softeners, and flavoring. The key to producing good bubbles is the proper gum base. As a representative of Amurol Products Company put it, "Gum base is the part that puts the 'chew' in chewing gum and the 'bubble' into bubble gum." Until recently, the gum base consisted mostly of tree resin; now, most manufacturers use polyvinyl acetate, a synthetic resin.

In order to produce a substantial bubble, the gum must be strong enough to withstand the pressure of the tongue and the formation of an air pocket but also flexible enough to stretch evenly as it expands. The secret ingredient in bubble gum is a class of ingredients called "plasticizers," a synthetic gum base that stretches farther than plain resin. Plasticizers guarantee sufficient elasticity to insure that little kids can pop bubbles big enough to plaster pink crud all over their chins and eyes simultaneously.

Submitted by Karin Norris of Salinas, California.

DAVID FELDMAN

Why Does Bazooka Joe Wear an Eye Patch?

Rest easily. Bazooka Joe has 20-20 vision and no eye deformity. But ever since he was introduced in 1953, Joe has donned an eye patch to give himself a little bit of that Hathaway Man panache.

And before you ask—Herman has always hidden behind his turtleneck, but he does have a perfectly functional neck.

Submitted by Christopher Valeri of East Northport, Rhode Island.

What Is the White Stuff on Baseball Card Bubble Gum and Why Is It There?

The white stuff is powdered sugar. And according to Bill O'Connor, of Topps, it is sprinkled on gum to keep it from sticking to other pieces of gum during the manufacturing process.

Both before and after "baseball" gum is cut to its final size, it is placed in stacks in a magazine. The powdered sugar prevents the pieces from clumping together. Bazooka brand gum, also made by Topps, doesn't need the powdered sugar because it isn't stacked in the same way.

Card gum contains less water than conventional bubble gum, but in humid conditions it absorbs moisture. The powdered sugar also prevents the wrapper from sticking to the gum on hot, sticky days.

DAVID FELDMAN

Why Are Baseball Card Wrappers Covered with Wax?

Wax wrap allows the cards and gum to be sealed with heat, an economical, quick, and safe method to secure the integrity of the packaging. But most bubble gum manufacturers are switching to poly-wraps because new equipment is faster and poly creates a more airtight seal. Now that baseball cards can be worth as much as objets d'art, it seems appropriate that the waxy texture of the wrappers will be eradicated.

Submitted by Kim Chase of Crestline, California.

Why Do Some Binoculars Have an Adjustment Only for the Right Eyepiece?

Militant left-handers swear to us that we live in a right-handers' world. But their argument doesn't hold up too well when it comes to binoculars; this is one case where the lefties have priority over those right-dominant types.

Binoculars can be focused in two ways. The "individual focus binocular" provides diopter scales for each eyepiece and spiral-type adjustments so that you can fix each eyepiece.

But our Imponderable refers to the "central focusing system," which has a focusing wheel in between the barrels of the two eyepieces. According to Bill Shoenleber, of Edmund Scientific Co.,

> This model is equipped with an individual diopter focus on *one* of the eyepieces (usually the right one). The center focus is used

DAVID FELDMAN

until the image seen by the left eye is clear. Then the diopter adjustment is used to adjust the focus for the right eye. Once corrected for your own individual diopter difference between eyes, it is then necessary only to use the center focus itself to get equally clear images for both eyes.

Bushnell and many other companies do make binoculars with the individual focus on the left, but for unknown reasons this configuration has never sold as well.

Submitted by Owen Elliott of Juno Beach, Florida.

Why Do Scabs Always Itch So Much?

Scabs don't itch, Ruth. People do.

Honest. Scabs are just crusts of dried blood and fiber that cover a wound. It's the wound that itches.

In *Imponderables,* we discussed how the itching sensation is sent through the same neural pathways as pain signals. In fact, most scientists and doctors believe that, as dermatologist Jerome S. Litt describes it, "An itch is a minuscule pain." Litt explains why the wound itches:

> In the healing process, some of the nerve fibers that mediate both pain and itch become irritated and inflamed. This process leads to the small pain (itch) we encounter. . . . Were these scabs deeper, we would then experience frank pain.

What happens during the recovery period to irritate nerve fibers? Wounds repair themselves and shrink in size, partly because of the elasticity of the skin, but partly because the scab pulls on the wound.

Less frequently, itching can be caused by infections or small cracks in the scab as it dries. Dermatologist Samuel T. Selden, of Chesapeake, Virginia, treats wounds with moist dressings, allowing the wound to heal without scabbing, and reports that he has not heard any complaints from patients about itching.

Submitted by Ruth Gudz of Prescott, Arizona. Thanks also to Tricia Roland of St. Louis, Missouri.

What Is the Purpose of the Holes on the Sides of Men's Hats? Decoration? Ventilation? A Receptacle for Feathers? Or?

Clothing historian and writer G. Bruce Boyer is emphatic: "The holes in the sides of men's hats are specifically and exclusively for ventilation."

Every hat manufacturer we spoke to agreed with Mr. Boyer. Feathers are usually placed in the hat band, not the holes. Nobody thought that the holes added much to the look of the hat.

So we went back to the poser of this Imponderable, proud of our newfound knowledge. And then he gave us a discomfiting response. If the holes are for ventilation, why does the sweat band inside of his hat cover the holes from the inside?

Hmmm.

Submitted by Ron Weinstock of New York, New York.

How Do They Peel and Clean Baby Shrimp?

Increasingly, by machine. The Laitram Corporation, based in New Orleans, Louisiana, dominates the field of automatic shrimp processing. With four separate stations, Laitram machines can process hundreds of pounds of shrimp per hour.

1. The high-capacity shrimp peeler can peel between 500 and 900 pounds of shells-on shrimp per hour, with or without heads.

2. The cleaner detaches unwanted gristle and waste appendages and then sends the shrimp on a flume ride to clean the crustaceans.

3. The waste separator segregates the waste material detached in the last step.

4. The deveiner deveins the shrimp.

These machines are neither sleek nor pretty—one peeler weighs more than two tons—but they save money. Machines also can grade shrimp and separate them by size, and they work just as easily on baby shrimp as jumbos.

HOW Do Football Officials Measure First Down Yardage with Chains, Especially When They Go on Field to Confirm First Downs?

In professional football, careers and millions of dollars can rest on a matter of inches. We've never quite figured out how football officials can spot the ball accurately when a running back dives atop a group of ten hulking linemen, let alone how the chain crew retains the proper spot on the sidelines and then carries the chain back out to the field without losing its bearings. Is the aura of pinpoint measurement merely a ruse?

Not really. The answer to this Imponderable focuses on the importance of an inexpensive metal clip. The National Football League's Art McNally explains:

> If at the start of a series the ball was placed on the 23-yard line in the middle of the field, the head linesman would back up

to the sideline and, after sighting the line of the ball, would indicate to a member of the chain crew that he wanted the back end of the down markers to be set at the 23-yard line. Obviously, a second member of the chain crew would stretch the forward stake to the 33-yard line.

Before the next down is run, one of the members of the chain crew would take a special clip and place that on the chain at the back end of the 25-yard line. In other words, the clip is placed on the five-yard marker that is closest to the original location of the ball.

When a measurement is about to be made, the head linesman picks up the chain from the 25-yard line and the men holding the front end of the stakes all proceed onto the field. The head linesman places the clip on the back end of the 25-yard line. The front stake is extended to its maximum and the referee makes the decision as to whether or not the ball has extended beyond the forward stake.

Thus the chain crew, when it runs onto the field, doesn't have to find the exact spot near the 23-yard line where the ball was originally spotted, but merely the 25-yard line. The clip "finds" the spot near the 23-yard line.

Submitted by Dennis Stucky of San Diego, California.

HOW Did Dr Pepper Get Its Name? Was There Ever a Real Dr. Pepper?

Yes, there was a real one, although he had a period after the "r" in "Dr". Dr. Charles Pepper owned a drug store in Rural Retreat, Virginia, and employed a young pharmacist named Wade Morrison.

Unfortunately for Wade, Dr. Pepper wasn't too happy when a romance blossomed between the young pharmacist and his attractive daughter. Pepper nixed the relationship, and the dejected Morrison moved to Waco, Texas, and opened Morrison's Old Corner Drug Store.

Morrison hired Charles Alderton, a young English pharmacist, whose duties included tending the store's soda fountain. Alderton noted the waning interest of his customers in the usual fruit-flavored soft drinks and decided to blend several fruit flavors himself. Alderton finally hit upon a concoction that satisfied Morrison and his taste buds.

Word of mouth spread about Alderton's new creation, and in 1885, what we now know as Dr Pepper became a popular item at the Corner Drug Store. But what would they call the new drink? The Dr Pepper Company supplied the answer:

> Morrison never forgot his thwarted romance and often spoke fondly of Dr. Pepper's daughter. Patrons of his soda fountain heard of the affair, and one of them jokingly suggested naming their new fountain drink after the Virginia doctor, thinking it would gain his favor. The new drink became known as Dr Pepper. It gained such widespread favor that other soda fountain operators in Waco began buying the syrup from Morrison and serving it.

Even certified Peppers might not realize that Dr Pepper is the oldest major soft-drink brand and was introduced to a national constituency at the 1904 World's Fair Exposition in St. Louis, a conclave which was to junk food what Woodstock was to the musical counterculture. (The St. Louis Fair also featured the debut of the ice cream cone, as well as hamburgers and hot dogs served in buns.)

Morrison made a fortune and in that sense wreaked some revenge on the real Dr. Pepper, but he never regained the attentions of Miss Pepper. Alderton, the actual originator of the drink, was content to mix pharmaceutical compounds, and was never involved in the operation of the Dr Pepper Company.

Submitted by Barth Richards of Naperville, Illinois. Thanks also to Kevin Hogan of Hartland, Michigan, and Josh Gibson of Silver Spring, Maryland.

DAVID FELDMAN

Why Is the Home Plate in Baseball Such a Weird Shape?

Until 1900, home plate was square like all the other bases. But in 1900, the current five-sided plate was introduced to aid umpires in calling balls and strikes. Umpires found it easier to spot the location of the ball when the plate was elongated. If you ask most players, it hasn't helped much.

Submitted by Bill Lachapell of Trenton, Michigan. Thanks also to Michael Gempe of Elmhurst, Illinois, and John H. McElroy of Haines City, Florida.

Why Do Hospital Gowns Tie at the Back?

It's bad enough being laid up in the hospital. Why do patients have to undergo the indignity of having their backsides exposed to all? This is a fashion statement that even sick people don't want to make.

We realize that hospital gowns aren't the first priority of hospital administrators, that items like nursing staffs, research budgets, and surgical care justifiably occupy much of their time. But while hospitals pursue the impossible dream of serving edible food, the eradication of the back-tied gown is possible right now.

The original justification for the back closure of hospital gowns was that this configuration enabled health care workers to change the gowns of the bedridden without disturbing the patients. If the gown tied in the front, the patients would have to be picked up (or lift themselves up) to remove the garment.

Perhaps it is growing concern for "patient modesty" (buzz words in the "patient apparel" industry), or perhaps it is jock-

DAVID FELDMAN

eying for competitive advantage among hospitals, especially private, for-profit hospitals, but many health-care administrators are starting to recognize the existence of alternatives to the back-tied gown. Scott Hlavaty, director of patient/surgical product management of uniform giant Angelica, told *Imponderables* that ties on the sides provide a maximum of patient modesty while requiring no more patient inconvenience to remove.

Angelica and other companies manufacture gowns for patients that minimize patient exposure and inconvenience when procedures are performed. Hlavaty explains:

> There are specialized gowns with "I.V. sleeves" that allow the gown to be removed by unsnapping the sleeves so that the I.V. tubes do not have to be removed from the patient in order to change a gown. Also, with the advent of pacemakers and heart monitors, "telemetry pockets" have been placed in the center of gowns. These pockets have openings in the back to allow for the pass through of the monitoring device so that these do not have to be disconnected either.

Some nurses we spoke to commented that gowns with back closures make it more convenient to give shots (in the backside, of course). But many patients prefer to wear their own pajamas, and nurses always manage to administer the shot.

Even if a patient were so incapacitated that a back closure was deemed best, many improvements have been made in hospital uniforms to prevent patients from exposing themselves to roommates and passersby. Back-closure gowns wouldn't be such a problem if the closures were made secure. Uniforms are now available with metal or plastic grippers, as well as velcro. And just as important, gowns are available with a "full overlap back," which provides enough material to overlap more like a bathrobe than a traditional hospital gown. At least with a full overlap gown, you have a shot at covering your rump if the closure unfastens.

Sure, these improvements in gown design cost a little more. But in times when a day in the hospital costs more than the weekly salary of the average person, who cares about a few more cents? After all, how can you encourage postsurgical patients to take a stroll around the hospital corridors when they're more concerned about being the objects of peeping Toms than they are about aches and pains?

Submitted by Diane M. Rhodes of Herndon, Virginia.

What's the Difference Between White Chocolate and Brown Chocolate?

One big difference seems to be that white chocolate doesn't exist. We were shocked to consult five dictionaries and find that none of them has a listing for "white chocolate." And the Food and Drug Administration, which regulates all the ingredients, properties, and definitions of chocolate, also does not recognize the existence of white chocolate.

Therefore, we may conclude that white chocolate is not a form of chocolate at all. Charlotte H. Connelly, manager of consumer affairs for Whitman's Chocolates, wrote *Imponderables* that because there is no legal definition of white chocolate, manufacturers are "not restricted to the type or the amount of ingredients that are incorporated in the 'white chocolate' recipe."

In practice, however, there is only one difference between white and brown chocolate—brown chocolate contains cocoa powder. Richard T. O'Connell, president of the Chocolate Manufacturers Association of the United States of America, explains:

> The cocoa bean from whence chocolate comes is ground into a substance called chocolate liquor (nonalcoholic) and when placed under hydraulic pressure, it splits into two parts, one of cocoa butter and the other cocoa powder. In normal "brown" chocolate, the chocolate liquor is usually mixed with additional cocoa butter to get that "melt-in-your-mouth" flavor. In "white" chocolate (which is not a chocolate), cocoa butter is usually mixed with

DAVID FELDMAN

sugar. Cocoa butter is light tan in color and, therefore, the term "white" is given to it.

Submitted by Vivian Delduca of Berkeley Heights, New Jersey. Thanks also to Fay Gitman of Pottsville, Pennsylvania.

Why Aren't Large-Type Books as Big as They Used to Be?

Twenty-five years ago, it was easy to spot the large-print books in bookstores—they were the ones that had to be housed in oversized bookshelves. And if you bought a few copies, you needed a dolly to haul them away.

No more. In the past, publishers simply enlarged the "regular" book until the print was big enough for the visually impaired to read. Now, most publishers reset the type and keep the size of the pages identical to those of its "small-print" companion.

Obviously, if the print is bigger and the book size is the same, either the margins must be reduced or the page count must increase. In most cases, book designers can't steal enough white space to avoid increasing the page count of the large-print book. But in order to keep them from getting too bulky and expensive, large-print manufacturers use thinner paper that is sufficiently opaque so that the print doesn't bleed over to the other side of the page.

What Causes the Green-Tinged Potato Chips We Sometimes Find? Are They Safe to Eat?

Potatoes are supposed to grow underground. But occasionally a spud becomes a little more ambitious and sticks its head out. Nature punishes the potato by giving it a nasty sunburn.

But why do potatoes turn green rather than red? No, it's not out of envy. The green color is chlorophyll, the natural consequence of a growing plant being exposed to light. According to Beverly Holmes, a public relations representative of Frito-Lay, chip producers try to eliminate the greenies. But a few elude them:

> We store our potatoes in dark rooms and have "pickers" on our production lines who attempt to eliminate [green] chips as they move along on the conveyers because of their undesirable appearance. However, a few chips can make their way through the production process.

Is it harmful to eat a green-tinged chip? Not at all. Chlorophyll stains are as harmless as the green beer or green bagels peddled on St. Patrick's day, and chlorophyll contains no artificial ingredients.

Submitted by Dr. John Hardin of Greenfield, Indiana. Thanks also to Ed Hirschfield of Portage, Michigan.

Why Are Tortilla Chips So Much More Expensive Than Potato Chips?

In the supermarket, potatoes are more expensive than corn. And potato chips are less expensive than corn tortilla chips. Doesn't anything make sense anymore?

Faced with our whining, snack food representatives remain unbowed. Expense of raw food supplies isn't the only determinant of food costs, they explained with patience and a tinge of exasperation. It's the processing of tortilla chips that makes them more expensive.

Even before it is cooked, the corn must be soaked for hours prior to processing. And then the fun begins. Al Rickard, director of communication for the Snack Food Association, explains:

> To make a tortilla chip, a snack manufacturer must cook the corn, grind it into a corn flour, mix it into the proper consistency, and send it through a large machine that rolls the dough into large sheets and cuts out the tortilla chips. The chips are then baked and fried before moving to the seasoning and packaging operations.
>
> By contrast, potato chips are made by simply washing, peeling, and slicing potatoes, which then move through a continuous fryer before moving out to the seasoning and packaging operation. Tortilla chip manufacturing requires more equipment and more labor, so the final cost is higher.

Frito-Lay's Beverly Holmes mentions that tortilla chips vary more in price than potato chips. Frito-Lay's Doritos brand is priced higher than potato chips, but in many markets, "restaurant-style" tortilla chips have been introduced.

> Restaurant-style tortilla chips are often sold in large, clear bags. They tend to be larger in size and are made with less salt, oil, and seasoning since these chips tend to be eaten with dips and sauces.

Submitted by John Morgan of Brooklyn, New York.

Does the Moon Have Any Effect on Lakes or Ponds? If Not, Why Does It Only Seem to Affect Oceans' Tides? Why Don't Lakes Have Tides?

If there is any radio show that we fear appearing on, it's Ira Fistel's radio show in Los Angeles. Fistel, a lawyer by training, has an encyclopedic knowledge of history, railroad lore, sports, radio, and just about every other subject his audience questions him about, and is as likely as we are to answer an Imponderable from a caller. Fistel can make a "Jeopardy!" Tournament of Champions winner look like a know-nothing.

So when we received this Imponderable on his show and we proceeded to stare at each other and shrug our shoulders (not particularly compelling radio, we might add), we knew this was a true Imponderable. We vowed to find an answer for the next book (and then go back on Fistel's show and gloat about it).

Robert Burnham, senior editor of *Astronomy*, was generous enough to send a fascinating explanation:

> Even the biggest lakes are too small to have tides. Ponds or lakes (even large ones like the Great Lakes) have no tides because these bodies of water are raised all at once, along with the land underneath the lake, by the gravitational pull of the Moon. (The

solid Earth swells a maximum of about eighteen inches under the Moon's tidal pull, but the effect is imperceptible because we have nothing that isn't also moving by which to gauge the uplift.)

In addition, ponds and lakes are not openly connected to a larger supply of water located elsewhere on the globe, which could supply extra water to them to make a tidal bulge. The seas, on the other hand, have tides because the water in them can flow freely throughout the world's ocean basins . . .

On the side of Earth nearest the Moon, the Moon's gravity pulls sea water away from the planet, thus raising a bulge called high tide. At the same time on the other side of the planet, the Moon's gravity is pulling *Earth* away from the *water,* thus creating a second high-tide bulge.

Low tides occur in between because these are the regions from which water has drained to flow into the two high-tide bulges. (The Sun exerts a tidal effect of its own, but only 46 percent as strong as the Moon's.)

Some landlocked portions of the ocean—the Mediterranean or the Baltic—can mimic the tideless behavior of a lake, although for different reasons. The Mediterranean Sea, for example, has a tidal range measuring just a couple of inches because it is a basin with only a small inlet (the Strait of Gibraltar) connecting it to the global ocean. The Gibraltar Strait is both narrow and shallow, which prevents the rapid twice-a-day flow of immense volumes of water necessary to create a pronounced tide. Thus the rise and fall of the tide in the Atlantic attempts to fill or drain the Med, but the tidal bulge always moves on before very much water can pour in or out past Gibraltar.

Alan MacRobert, of *Sky & Telescope,* summarizes that a body of water needs a large area to slosh around in before tidal effects are substantial, and he provides a simple analogy:

Imagine a tray full of dirt dotted with thimbles of water, representing a land mass with lakes. You could tilt it slightly and nothing much would happen. Now imagine a tray full of water—an ocean. If you tilted it just a little, water would sloop out over your hands.

Submitted by a caller on the Ira Fistel show, KABC-AM, Los Angeles, California.

DO PENGUINS HAVE KNEES?

~AND YOUR CELESTIAL SECURITY NUMBER IS~?

Why Do the Backs of Social Security Cards Say "Do Not Laminate" When We Are Expected to Keep the Cards for Our Entire Lives?

The main purpose of the social security card seems to be to prove that we exist. And laminating a card hampers the ability of the government to ascertain whether a card has been tampered with or counterfeited. If your card doesn't pass muster, you don't exist, so we're talking about important stuff here.

John Clark, officer of the Social Security Administration, told *Imponderables* that the social security card incorporates several security features to foil would-be card defacers:

- The [card] stock contains a blue tint marbleized random pattern. Any attempt to erase or remove data is easily detectable because the tint is erasable.
- Planchets (small multi-colored discs) are randomly placed on the paper stock and can be seen with the naked eye.
- Intaglio printing of the type used in U.S. currency is used for

some printing on the card and provides a raised effect that can be felt.

"A laminated card hampers the ability of the government to utilize these security features," Clark summarizes.

Sure, the Social Security Administration would love us to keep the same social security card until we die, but it is used to doling out replacements for lost or damaged cards.

And there's good news to report. Sure, the government won't let you laminate your social security card, but it will replace it for free. Trish Butler, associate commissioner for public affairs for the Social Security Administration, asked us to remind *Imponderables* readers that "there is *never* a charge for any service we provide."

Now if only the IRS would adopt the same policy, we'd be happy campers.

Submitted by Kristi Nelson of Vancouver, Washington. Thanks also to April Pedersen of Edmond, Oklahoma.

Why Are Nonsweet Wines Called "Dry"?

"Sweet" makes sense. Sweet wines *do* have more sugar in them than dry ones. The main purpose of the sugar is to combat the acidity of the tannic and other acids found in wine.

Consumers may disagree sharply about how much sugar they prefer in wines, but can't we all agree that "dry" wine is just as wet as sweet wine?

Surprisingly, few of our wine experts could make any sense of "dry" either, but two theories emerged. Spirits expert W. Ray Hyde argues that the terminology stems from both the sensory experience of tasting and more than a little marketing savvy:

Sugar stimulates the saliva glands and leaves the mouth wet. Acids, on the other hand, have an astringent quality that leaves the mouth feeling dry. Winemakers know that the consumer prefers a "sweet" wine to a "wet" wine and a "dry" wine to an "acidic" wine.

But Irving Smith Kogan, of the Champagne News and Information Bureau, wrote *Imponderables* about an intriguing linguistic theory:

> . . . the explanation is in the French language. "*Sec*" is a synonym of lean, and means *peu charnu* (without flesh), without softness or mellowness. This image appears in the English expression "bone-dry." "*Sec*" also means neat, as in undiluted, pure, bare, raw ("*brut*" in French), i.e., unsweetened.
>
> The issue of "dry" versus "sweet" is not the same for Champagne as for still wines. In the case of Champagne, the wine was originally labeled "*doux*" which is the French word for sweet. But in the mid-nineteenth century a Champagne-maker named Louise Pommery decided to make a less-sweet blend and called it "*demi-sec*" (half-dry), which is still quite sweet but less so than the *doux*.
>
> Since her day, Champagnes have been blended progressively dryer (i.e., less and less sweet). So, today we have a range of Champagnes in ascending order of dryness, demi-sec, sec, extra-dry, brut, and extra-brut. The doux is no longer commercially available.

Kogan adds that the above etymology of "dry" does not apply to still wines, for which "dry" simply means not sweet. Notice our current bias for dry champagne. Now the "driest" champagne you can buy is half-bone-dry.

Submitted by Bob Weisblut of Wheaton, Maryland.

Why Do the Rear Windows of Taxicabs (and Some Other Cars) Not Go Down All the Way?

Although we associate this Imponderable with cabs, everyone we spoke to in the taxi industry assured us that they didn't modify the back windows of their fleet cars. Nor was the movement of back seat windows of any concern when considering which models to buy (although it can be of great concern to passengers —when is the last time you've ridden in an air-conditioned cab?). They provide the rear windows for their fares that Detroit (or Japan) provides them.

So the real question is why auto manufacturers don't design their back windows to go down all the way. C. R. Cheney, of Chrysler's Engineering Information Services department, wrote to us about this decision:

> With the advent of automotive air conditioning, the need for this feature disappeared, since it was no longer necessary to pro-

mote maximum flow-through ventilation in this way (at one time, windshields could be opened, too). We can also probably make a good case for improved safety in vehicles with a fixed rear window because this area can no longer be a path of entry for exhaust fumes, insects that could distract the driver, noise, etc.

Of course, occupants of a car are as worried about what might go *out* of a car as what might traipse in. Max E. Rumbaugh, Jr., vice-president of the Society of Automotive Engineers, wrote that

> Some engineers in the past have been known to limit the downward movement of rear windows in the belief that customers want protection that prohibits young children from climbing out of a wide open rear window while a vehicle is moving down a highway.

And all things being equal, why wouldn't manufacturers design a rear window so that customers could put the window down as far as they want? Rumbaugh explains:

> . . . some engineers may be faced with constraints caused by the design and manufacture of a smaller car. In smaller cars, the location of the rear wheel dictates the shape of the rear door. This shape can force a restriction on the downward movement of a full-size rear window.

Submitted by Joanne Walker of Ashland, Massachusetts. Thanks also to David A. Kroffe of Los Alamitos, California; Stephanie Suits of San Jose, California; and Renee Tribitt of Minot, North Dakota.

What Is the Two-Tone Signal at the Start of Many Rented Videotapes?

Yes, it has a purpose other than to puncture your eardrums. Or at least that's what several videotape experts have claimed, anyway.

DAVID FELDMAN

William J. Goffi, of Maxell Corporation, told *Imponderables* that the tone

> is used to facilitate the recording and loading process of video-tapes. As the videotapes are duplicated on high-speed duplicators, the tone "tells" the machine to either stop or start the duplicating process. As the tape is loaded into the shells, the tone again "tells" the loader where the tape starts and ends.

Not all professional duplication is done on high-speed machines. Some production houses duplicate in real time, with one "master" machine supplying the material for dozens of "slave" copying machines. According to Panasonic's public relations manager, Mark Johnson, the two-tone signal is also used to set the audio levels for the slave machines.

All we know is that the two-tone signal sure doesn't help us set the proper volume level on our television. Nothing can sell us on the idea of a remote that can set volume levels more than a trip up to the TV set to turn down the shrieking blast of the two-tone frequency (which only a dog could love). Inevitably, we follow up with a weary trek, fifteen seconds later, to turn up the sound so that our witty comedy doesn't turn out to be an unintentional mime show.

Submitted by Cary Chapman of Homeland, California.

How Was 911 Chosen as the Uniform Emergency Telephone Number?

Old-timers like us will remember when the codes for telephone services were not uniform from city to city. In one town, "information" could be found at 411; in another, at 113. The Bell system needed to change this haphazard approach for two reasons. Making numbers uniform throughout the country would promote ease of use of their services. And reclaiming the 1 as an access code paved the way for direct dialing of long-distance calls.

Most of AT&T Bell's service codes end in 11. 211, 311, 411, 511, 611, 711, and 811 were already assigned when pressure accumulated to create a uniform, national number for emergencies. According to Barbara Sweeney, researcher at AT&T Library Network Archives, all the numbers up to 911 had already been assigned. So 911 become the emergency number by default.

Think of how sophisticated the automated routers of the phone system are. When you dial 411 for directory assistance, each digit is crucial in routing the call properly. The first digit, 4, tells the equipment that you are not trying to obtain an operator ("0") or make a long-distance call. The second and third digits, 11, could be used in an area code as well as an office code, so the equipment has to be programmed to recognize 211, 311, 411, 511, 611, 711, 811, and 911 as separate service codes and not "wait" for you to dial extra digits before connecting you with the disconnected recording that will tell you the phone number of Acme Pizza.

Submitted by Karen Riddick of Dresden, Tennessee.

Why Do Birds Bother Flying Back North After Migrating to the South?

Why bother flying so many miles south, to more pleasant and warmer climes, only to then turn around and trudge back to the only seasonably hospitable northeastern United States? Come to think of it, are we talking about birds here or half the population of Miami?

We're not sure what motivates humans to migrate, but we do have a good idea of what motivates birds to bother flying back north again. Of course, birds probably don't sit around (even with one leg tucked up in their feathers) thinking about why they migrate; undoubtedly, hormonal changes caused by natural breeding cycles trigger the migration patterns. After speaking to several bird experts, we found a consensus on the following reasons why birds fly back to the North:

>   1. *Food.* Birds fly back north to nest. Baby birds, like baby humans, are ravenous eaters and not shy about demanding food.

DO PENGUINS HAVE KNEES?

As Todd Culver of the Laboratory of Ornithology at Cornell University put it, "The most likely reason they return is the super abundant supply of insects available to feed their young." The more food the parents can raise, the healthier the offspring will be and the lower the babies' mortality rate.

2. *Longer daylight hours.* The higher the latitude, the longer period of daylight parents have to find food and feed it to their babies. Some birds find food sources solely by using their vision; they cannot forage with any effectiveness in the dark.

3. *Less competition for food and nesting sites.* If all birds converged in the southern latitudes when nesting, it would be as easy to find a peaceful nest and plentiful food sources as it would be to find a quiet, pleasant, little motel room in Fort Lauderdale during spring break.

4. *Safety.* Birds are more vulnerable to predators when nesting. In most cases, there are fewer mammal predators in the North than in the South. Why? Many mammals, who don't migrate, can't live in the North because of the cold weather.

5. *Improved weather.* Some birds migrate south primarily to flee cold weather in the North. If they time it right, birds come back just when the weather turns pleasant in the spring, just like those humans in Florida.

Submitted by Michael S. Littman of Piscataway, New Jersey. Thanks also to Jack Weber of Modesto, California; Lori Tomlinson of Newmarket, Ontario; and Saxon Swearingen of La Porte, Texas.

DAVID FELDMAN

Why Are the Oceans Salty? What Keeps the Oceans at the Same Level of Saltiness?

Most of the salt in the ocean is there because of the processes of dissolving and leaching from the solid earth over hundreds of millions of years, according to Dr. Eugene C. LaFond, president of LaFond Oceanic Consultants. Rivers take the salt out of rocks and carry them into oceans; these eroded rocks supply the largest portion of salt in the ocean.

But other natural phenomena contribute to the mineral load in the oceans. Salty volcanic rock washes into them. Volcanos also release salty "juvenile water," water that has never existed before in the form of liquid. Fresh basalt flows up from a giant rift that runs through all the oceans' basins.

With all of these processes dumping salt into the oceans, one might think that the seas would get saturated with sodium chloride, for oceans, like any other body of water, keep evaporating. Ocean spray is continuously released into the air; and the recycled rain fills the rivers, which aids in the leaching of salt from rocks.

Yet, according to the Sea Secrets Information Services of the International Oceanographic Foundation at the University of Miami, the concentration of salts in the ocean has not changed for quite a while—about, oh, 1.5 billion years or so. So how do oceans rid themselves of some of the salt?

First of all, sodium chloride is extremely soluble, so it doesn't tend to get concentrated in certain sections of the ocean. The surface area of the oceans is so large (particularly since all the major oceans are interconnected) that the salt is relatively evenly distributed. Second, some of the ions in the salt leave with the sea spray. Third, some of the salt disappears as adsorbates, in the form of gas liquids sticking to particulate matter that sinks below the surface of the ocean. The fourth and most dramatic way sodium chloride is removed from the ocean is by the large accumulations left in salt flats on ocean coasts, where the water is shallow enough to evaporate.

It has taken so long for the salt to accumulate in the oceans that the amount of salt added and subtracted at any particular time is relatively small. While the amount of other minerals in the ocean has changed dramatically, the level of salt in the ocean, approximately 3.5 percent, remains constant.

Submitted by Merilee Roy of Bradford, Massachusetts. Thanks also to Nicole Chastrette of New York, New York; Bob and Elaine Juhre of Kettle Falls, Washington; John H. Herman of Beaverton, Oregon; Matthew Anderson of Forked River, New Jersey; and Cindy Raymond of Vincentown, New Jersey.

DAVID FELDMAN

How Do 3-D Movies and 3-D Glasses Work?

3-D movies are a variation of the stereovision systems (e.g., Viewmasters) that we see used in tourist trinkets and children's toys. These devices present two different views a few inches apart from the viewpoint of the human eyes. The left image is presented only to the left eye and the right image is sent directly to the right eye.

But the technology for a 3-D film is more complicated, because the filmmaker must invent some way to keep the left eye from seeing what only the right eye is supposed to view, even though both images are being projected on the screen simultaneously. The history of the technology was reviewed for us by David A. Gibson, of the Photo Equipment Museum of Eastman Kodak Company:

> The first system was invented in the 1890s, and the images are called anaglyphs. The left-eye image was projected with a red colored filter over the projector lens and a blue-green filter [was

put] over the lens of the projector for the right-eye image. Glasses with the same color filters were used in viewing the images—the red filter for the left eye transmitted the light from the left-image projector, and blocked the light from the right-image projector. This system has also been used to print such things as stereo comic books and has been used experimentally with stereo images broadcast on color television.

The only problem with this technology is that it works best with monochrome images. The red and blue-green tints of the glasses add unwanted and unsubtle coloration to a color 3-D film.

The solution, Gibson is generous enough to admit, came from rival Polaroid, which developed, appropriately enough, a polarization method specifically for 3-D films. The Polaroid technology beams

> the angle of polarization for one eye at right angles to that for the other eye, so that one image is transmitted while the other eye is blocked. This is the system used for the 3-D movies made in the 1950s.

Submitted by Don Borchert of Lomita, California.

What Use Is the Appendix to Us? What Use Are Our Tonsils to Us?

The fact that this Imponderable was first posed to us by a medical doctor indicates that the answer is far from obvious. We asked Dr. Liberato John DiDio this question, and he called the appendix the "tonsils of the intestines." We wondered if this meant merely that the appendix is the organ in the stomach most likely to be extracted by a surgeon. What good are organs like tonsils and the appendix and gall bladder when we don't seem to miss them at all once they've been extracted?

Actually, tonsils and the appendix do have much in common. They are both lymphoid organs that manufacture white blood cells. William P. Jollie, professor and chairman of the Department of Anatomy at the Medical College of Virginia at Virginia Commonwealth University, explains the potential importance of the appendix:

> One type of white blood cell is the lymphocyte; it produces antibodies, proteins that distinguish between our own body proteins and foreign proteins, called antigens. Antibodies, produced by lymphocytes, deactivate antigens.
>
> Lymphocytes come in two types: B-lymphocytes and T-lymphocytes. T-lymphocytes originate in the thymus. There is some evidence to suggest that B-lymphocytes originate in the appendix, although there is also evidence that bone marrow serves this purpose.

If our appendix is so important in fighting infections, how can we sustain its loss? Luckily, other organs, such as the spleen, also manufacture sufficient white blood cells to take up the slack.

Some doctors, including Dr. DiDio, even suggest that the purpose of the appendix (and the tonsils) might be to serve as a lightning rod and actually attract infections. By doing so, the theory goes, infections are localized in one spot that isn't critically important to the functioning of the body. This lightning-rod theory is supported, of course, by the sheer numbers of people who encounter problems with appendix and tonsils compared to surrounding organs.

Accounts vary on whether patients with extracted tonsils and/or appendix are any worse off than those lugging them around. It seems that medical opinion on whether it is proper to extract tonsils for mild cases of tonsillitis in children varies as much and as often as hemlines on women's skirts. Patients in the throes of an appendicitis attack do not have the luxury of contemplation.

Submitted by Dr. Emil S. Dickstein of Youngstown, Ohio.
Thanks also to Lily Whelan of Providence, Rhode Island, and
George Hill of Brockville, Ontario.

Where Does the Moisture Go When Wisps of Clouds Disappear in Front of Your Eyes?

A few facts about clouds will give us the tools to answer this question:

1. A warm volume of air at saturation (i.e., 100 percent relative humidity), given the same barometric pressure, will hold more water vapor than a cold volume of air. For example, at 86 degrees Fahrenheit, seven times as much water vapor can be retained as at 32 degrees Fahrenheit.

2. Therefore, when a volume of air cools, its relative humidity increases until it reaches 100 percent relative humidity. This point is called the dew point temperature.

3. When air at dew point temperature is cooled even further, a visible cloud results (and ultimately, precipitation).

4. Therefore, the disappearance of a cloud is caused by the opposite of #3. Raymond E. Falconer, of the Atmospheric Sciences Research Center, explains:

> As a volume of air moves downward from lower to higher barometric pressure, it becomes warmer and drier, with lower relative humidity. This causes the cloud to evaporate.
>
> When we see clouds, the air has been rising and cooling with condensation of the invisible water vapor into visible cloud as the air reaches the temperature of the dew point. When a cloud encounters drier air, the cloud droplets evaporate into the drier air, which can hold more water vapor.
>
> When air is forced up over a mountain, it is cooled, and in the process a cloud may form over the higher elevations. However, as the air descends on the lee side of the mountain, the air warms up and dries out, causing the cloud to dissipate. Such a cloud formation is called an orographic cloud.

Submitted by Rev. David Scott of Rochester, New York.

DAVID FELDMAN

Why Is Frozen Orange Juice Just About the Only Frozen Product That Is Cheaper Than Its Fresh Counterpart?

Our correspondents reasoned, logically enough, that since both fresh and frozen orange juice are squeezed from oranges, and since the frozen juice must be concentrated, the extra processing involved would make the frozen style more expensive to produce and thus costlier at the retail level. But according to economists at the Florida Department of Citrus, it just ain't so.

Although the prices of fresh fruit, chilled juice, and frozen concentrate are similar at the wholesale level (within two cents of the fresh fruit equivalent per pound), they depart radically at the retail level. In the period of 1987–1988, frozen concentrated orange juice cost 24.6 cents per fresh pound equivalent; chilled juice was 35 cents per fresh pound equivalent; and the fruit itself a comparatively hefty 58.5 cents per pound. Why the discrepancy?

In the immortal words of real estate brokers across the world, the answer is: location, location, location. Chilled juice (and the fresh fruit itself, for that matter) would cost less than frozen if it didn't need to be shipped long distances. But it does. The conclusion, as Catherine A. Clay, information specialist at the Florida Department of Citrus, elucidates, is clearly that the costs at the retail level are due to distribution rather than processing costs.

> One 90-pound box of oranges will make about 45 pounds of juice. Concentrating that juice by removing the water will reduce the weight by at least two-thirds, so the amount of frozen concentrate in one box would be about 15 pounds.
>
> So you can ship three times as much frozen concentrated juice in one truck than you can chilled juice, and twice as much chilled juice as fresh oranges. In the space in which one 90-pound box of oranges is shipped, you could ship six times as much frozen concentrated orange juice.

In addition, fresh fruit can begin to decay or be damaged during transit to the retailer. So while the retailer may have paid for the entire truckload, he may have to discard decayed or damaged fruit. Yet his cost for that load remains the same regardless of how much actual fruit he sells.

Frozen concentrated juice does not spoil, so there is no loss. The higher price for fresh fruit would compensate the retail buyer for the cost of the lost fruit.

Clay didn't add that many juice distributors who use strictly Florida oranges for their chilled juice use cheaper, non-American orange juice for their concentrated product. Why don't they use the cheaper oranges for their chilled juices? Location. Location. Location. It would be too expensive to ship the heavier, naturally water-laden fruits thousands of miles.

Submitted by Eugene Hokanson of Bellevue, Washington.
Thanks also to Herbert Kraut of Forest Hills, New York.

DAVID FELDMAN

MAGAZINES: LIGHT READING, FOLLOWED BY **HEAVY CLEANING!**

HOW Are the Subscription Insert Cards Placed in Magazines? What Keeps Them from Falling Out As the Magazine Is Sent Through the Postal System?

Fewer things are more annoying to us than receiving a magazine we put our soft-earned bucks down to subscribe to, and being rewarded for our loyalty by being showered with cards entreating us to subscribe to the very magazine we've just shelled out for.

Why is it necessary to have to clean a magazine of foreign matter before you read the darn thing? Because the cards work. Publishers know that readers hate them; but the response rate to a card, particularly one that allows a free-postage response, attracts more subscribers than a discreet ad in the body of the magazine.

Those pesky little inserts that fall out are called "blow-in cards" in the magazine biz. We thought "blow-out" cards might be a more descriptive moniker until we learned the derivation

DO PENGUINS HAVE KNEES? 157

of the term from Bob Nichter, of the Fulfillment Management Association.

Originally, blow-in cards were literally blown into the magazine by a fan on the printer assembly line. Now, blow-ins are placed mechanically by an insertion machine after the magazine is bound. Nothing special is needed to keep the cards inside the magazine; they are placed close enough to the binding so that they won't fall out unless the pages are riffled.

Why do many periodicals place two or more blow-in cards in one magazine? It's not an accident. Most magazines find that two blow-in cards attract a greater subscription rate than one. Any more, and reader ire starts to overshadow the slight financial gains.

Submitted by Curtis Kelly of Chicago, Illinois.

Why Do We Put Thermometers Under Our Tongues? Would It Make Any Difference If We Put Them Above Our Tongues If Our Mouths Were Closed?

Anyone who has ever seen a child fidgeting, desperately struggling to keep a thermometer under the tongue, has probably wondered why this practice started. Why do physicians want to take our temperature in the most inconvenient places?

No, there is nothing intrinsically important about the temperature under the tongue or, for that matter, in your rectum. The goal is to determine the "core temperature," the temperature of the interior of the body.

The rectum and tongue are the most accessible areas of the body that are at core temperature. Occasionally the armpit will be used, but the armpit is more exposed to the ambient air, and tends to give colder readings. Of course, drinking a hot beverage, as many schoolchildren have learned, is effective in shooting one's temperature up. But barring tricks, the area under the

DAVID FELDMAN

tongue, full of blood vessels, is almost as accurate as the rectal area, and a lot more pleasant place to use.

So what are the advantages of putting the thermometer *under* the tongue as opposed to over it? Let us count the ways:

1. *Accuracy.* Placing the thermometer under the tongue insulates the area from outside influence, such as air and food. As Dr. E. Wilson Griffin III told *Imponderables,* "Moving air would evaporate moisture in the mouth and on the thermometer and falsely lower the temperature. It is important to have the thermometer under the tongue rather than just banging around loose inside the mouth, because a mercury thermometer responds most accurately to the temperature of liquids or solids in direct contact with it. . . ."

2. *Speed.* The soft tissues and blood vessels of the tongue are ideal resting spots for a thermometer. Dr. Frank Davidoff, of the American College of Physicians, points out that compared to the skin of the armpit, which is thick, horny, and nonvascular, the "soft, unprotected tissues under the tongue wrap tightly around the thermometer, improving the speed and completeness of heat transfer."

3. *Comfort.* Although you may not believe it, keeping the thermometer above the tongue would not be as comfortable. The hard thermometer, instead of being embraced by the soft tissue below the tongue, would inevitably scrape against the much harder tissues of the hard palate (the roof of your mouth). Something would have to give—and it wouldn't be the thermometer.

Davidoff concedes that in a pinch, placing the thermometer above the tongue might not be a total disaster:

In principle, you could get a reasonably accurate temperature reading with a thermometer above your tongue *if* you hadn't recently been mouth breathing or hadn't recently eaten or drunk anything, *if* you held the thermometer reasonably firmly between your tongue and the roof of your mouth, and *if* you kept it there long enough.

Do Penguins Have Knees?

They sure do, although they are discreetly hidden underneath their feathers. Anatomically, all birds' legs are pretty much alike, although the dimensions of individual bones vary a great deal among species.

Penguins, like other birds, have legs divided into three segments. The upper segment, the equivalent of our thigh, and the middle segment, the equivalent of our shinbone, or the drumstick of a chicken, are both quite short in penguins.

When we see flamingos, or other birds with long legs, they appear to possess a knee turned backwards, but these are not the equivalent of a human knee. Penguins, flamingos, and other birds do have knees, with patellas (knee caps) that bend and function much like their human counterparts.

We spoke to Dr. Don Bruning, curator of ornithology at the New York Zoological Park (better known as the Bronx Zoo), who told us that the backwards joint that we perceive as a knee in flamingos actually separates the bird equivalent of the ankle from the bones of the upper foot. The area below the backwards joint is not the lower leg but the upper areas of the foot. In other words, penguins (and other birds) stand on their toes, like ballet dancers.

Penguins are birds, of course, but their element is water rather than sky. Penguins may waddle on land, but their legs help make them swimming machines. Penguins use their wings as propellers in the water, and their elongated feet act as rudders.

So rest assured. Even if you can't see them, penguins have legs (with knees). And they know how to use them.

Submitted by John Vineyard of Plano, Texas. Thanks also to Ruth Vineyard of Plano, Texas.

HOW Do They Make Hot Dog Buns That Are Partially Sliced?

Now that we solved the Imponderable of why there are ten hot dogs in a package and only eight hot dog buns in a package (see *Why Do Clocks Run Clockwise?*), we can tackle a few less challenging bread Imponderables without guilt.

Barbara K. Rose, manager of consumer affairs at Continental Baking Company (the folks who bring us Wonder Bread and Twinkies), answered this question with ease:

> The hot dog buns are removed from the baking pans and placed on a conveyer-type system. These buns slide past circular blades that are mounted on rotating vertical shafts. Two buns are allowed between each blade: the bun on the right is sliced on the right side; the bun on the left is sliced by a blade on the left. These blades are set to slice only a specific distance into the bun and will not slice through. The tops of the buns hold them together.

Submitted by Laurie Hutler of Boulder Creek, California.
Thanks also to Robert Chell of Harrisonburg, Virginia, and Deb Graham of Mt. Vernon, Washington.

How Do They Fork Split English Muffins? What Causes the Ridges in English Muffins?

English muffins also run past circular blades, actually two blades, which slice only one-quarter inch or so into the muffin. But each muffin is also "forked," passed through two spinning wheels with Roman spear points. These spears penetrate into the muffin anywhere from one to one and one-half inch, depending upon the baker's preference.

Tom Lehmann, director of baking assistance at the American Institute of Baking, told *Imponderables* that the initial one-quarter-inch slice of the outer edges of the muffin yields a smooth consistency. The forking doesn't sever the muffin into two separate pieces but does produce the perforations by which they slice the muffin and the ridges, nooks, and crannies that provide the rough texture for which English muffins are famous.

Continental Baking's Barbara Rose reports that her company found that fork splitting didn't work on their Raisin Rounds, which are sliced all the way through the muffin: "Raisin Rounds must be sliced because the raisins will accumulate on the fork splitting machinery causing several mechanical breakdowns and halting production."

Submitted by Donna Burks of Gilroy, California. Thanks also to Ruth Mascari of Monkton, Maryland.

DAVID FELDMAN

Why Do Cat Hairs Tend to Stick to Our Clothes More Than Those of Dogs or Other Pets?

A cat's hair is the most electrostatic of all pet hairs, which may be the main reason why cat hairs stick to clothes. But the physiology of cat hair also contributes to kitty cling. Dr. John Saidla, assistant director of the Cornell Feline Health Center, explains:

> The hair coat in the cat consists of three different types of hair: primary or guard hairs within the outer coat: awn hairs (intermediate-sized hairs forming part of the primary coat); and secondary hairs (downy hair found in the undercoat).
>
> The guard hairs are slender and taper towards the tip. The cuticles on these hairs have microscopically small barbs that are very rough. This is the main reason cat hairs stick to clothing, and it is this hair that is found most commonly on clothing.
>
> The awn hairs have broken or cracked cuticles that are rough and would aid in their clinging, also. The secondary hairs are very

thin and are wavy or evenly crimped. These hairs are caught and held in more roughly textured clothing.

Submitted by an anonymous caller on Owen Spann's nationally syndicated radio show.

Why Are You Never Supposed to Touch a Halogen Light Bulb with Your Fingers?

Conventional light bulbs use soda-lime glass, which is perfectly functional. But tungsten-halogen bulbs are made of much more durable quartz glass because they must withstand much higher temperatures, a minimum of 250 degrees Centigrade.

Rubin Rivera, of Philips Lighting, told *Imponderables* that quartz halogen lamps must not be touched with the fingers because the natural oils from the skin, in combination with the high temperatures reached by the bulbs when illuminated, will cause the crystalline structure of the bulb glass to change.

Caden Zollo, product manager of The Specialty Bulb Co., adds that contact with human oils can cause the glass to crack and leak. Air can then get into the filament and, in extreme conditions, can cause the bulb not only to leak but to explode.

To avoid this "explosive" situation, some halogen lamps come with a separate outer bulb so that the lamps can be touched. If your hands have come in contact with the bulb, or you need to clean the bulb, wiping it with denatured alcohol will reverse the effect of your greasy hands.

Submitted by Gail Lee of Dallas, Texas.

Why Is There an "H" Inside of the "C" in the Hockey Uniform of the Montreal Canadiens?

The "H" stands for hockey. When the team was founded in 1909, it was known as "Club Canadien" and its sweaters sported a big white "C." This tradition lasted all of one year, when the club switched to a red uniform with a green maple leaf and a Gothic "C." Not content to rest on its fashion statement, the team changed its look again, adopting a blue, white, and red "barber pole" symbol featuring the letters "CAC," which stood for Club Athlétique Canadien.

In 1917, the Club Athlétique Canadien folded, but owner George Kennedy replaced the "A" inside the large "C" with the letter "H," to signify "hockey." The letters then stood for "Club de Hockey Canadien," the official name of the team for more than seventy years.

Submitted by Bob Tatu of Conshohocken, Pennsylvania.

Why Don't Radio Shack Stores Use Cash Registers?

Have you ever noticed that most of the time when you make a purchase at Radio Shack, salespeople ask you for your address? Clearly, the Tandy Corporation likes to compile as much information about its business as possible, and no figures are of more interest than how their individual stores are faring in sales and inventory control.

Befitting a retail business that specializes in high technology, cash registers don't impart enough information to satisfy the Tandy Corporation. Ed Juge, Radio Shack's director of market planning, explains:

> Radio Shack does not use cash registers. They're decades-old technology. Every company-owned Radio Shack store is

equipped with state-of-the-art electronic point-of-sale terminals, which are really "diskless" Tandy 1000 SX computers, tied to a multi-user Tandy computer in the office area.

These POS systems assure correct pricing and inventory counting of every item sold, along with a lot of other information that helps us run our business more efficiently. Every evening by 7:00 P.M., we can tell you exactly how many of any one of our 3000+ line items were sold that day in our 5,000 stores across the country.

Each individual store transmits sales data, every day, to Fort Worth [Tandy's corporate headquarters]. As that is being done, our Fort Worth computer can update the individual stores' files with new prices, or newly available product information, as well as sending them important information bulletins.

The system assures that every customer is getting the benefit of every sale price in effect on the date of purchase [we assume this up-to-the-minute price accuracy applies to price rises, as well], even if the sale price somehow escaped notice by our employees. Sales are written up about three times as quickly, and with the error rate reduced by a factor of one hundred to one over previous methods.

Don French, chief engineer for Radio Shack, adds that in many cases salespeople are away from the counter helping customers, making it relatively easy to have money stolen: "Keeping the money in a drawer makes it a little harder for this to happen."

Submitted by Doc Swan of Palmyra, New Jersey.

How Do They Decide Where to Put Thumbnotches in Dictionaries?

A frustrated David S. Clark wrote to us:

> When I updated my *New World Dictionary* a year or two ago I thought I'd bought a defective copy. The thumb indexes didn't match with the first page of the letter indicated. When I later bought a *Merriam-Webster's* and found the same situation, I concluded it must be a cheaper way to thumb index. Is there some other reason?

A good question, David, and you were right in concluding that saving money had a lot to do with the new indexing system. Maybe not too many other people have lost sleep over this issue, but dictionary publishers have.

Some form of visual indexing is necessary to make a dictionary user-friendly. But thumbnotches are expensive to install. Most unabridged dictionaries contain thumbnotches at the beginning of each alphabet entry, but this is impossible in thinner collegiate dictionaries, where twenty-six thumbnotches would bleed into each other. Furthermore, thumbnotching at the beginning of each letter is expensive.

Merriam-Webster Inc. found an elegant solution that has been widely imitated. John M. Morse, manager of editorial operations and planning, explains:

> For a good many years, thumbnotches in our dictionaries have referenced two letters of the alphabet each; for example, in the *Ninth Collegiate* there is one notch for *L* and *M*. In the past, this *LM* tab would have been placed at the beginning of the letter *L*, but in 1978 (during the life of the eighth edition) we changed over to an automatic thumbnotching system which places the tab somewhere close to the middle of the pages devoted to the referenced letters. In part this was done to control the costs of thumbnotching and so the price of the dictionary, but it is not really a disadvantage to the dictionary user.
>
> When a user goes to the dictionary to find a word beginning with *L* or *M*, we have no way of knowing which word is being sought, but we do know it will fall within the roughly 120 pages on which words beginning with *L* and *M* appear. Under the old system, one could find the tab and still be 100 pages or more away from the desired word. With the tab nearly in the middle of those pages, one is never more than, say, 70 pages away.
>
> It is true that this is a break with tradition and it is also true that we have received perhaps several hundred letters from users who feel uncomfortable with the new system, but we think most people become used to the arrangement quite quickly. We have sold millions of dictionaries indexed in this way, and other publishers have since followed our lead, so that now it is fair to say that most thumb-indexed desk dictionaries sold in this country use this indexing system.

Submitted by David S. Clark of Northridge, California.

DAVID FELDMAN

What Is the Purpose of the Button You Press to Unlock the Key from the Ignition on Some Cars? If It Is a Worthwhile Mechanism, Why Isn't It on All Cars?

The purpose of the ignition locking mechanism is to comply with the Federal Motor Vehicle Safety standard that mandates that the key that deactivates the engine also lock the steering column. But foiling a car thief poses safety risks. If you accidentally pulled out the key while the car was in motion, you had better pray for sparse traffic and rubber guard rails on the road: You wouldn't be able to steer.

To avoid inadvertent key withdrawal and the resultant steering column locking, the standard mandates that a separate action (other than turning the key) be required to pull the key out of the ignition. If the car's gear shift selector is mounted on the column, the movement to the park position is considered a "separate action."

But if the shift is mounted on the floor or console, another motion is required. Thomas J. Carr, director of safety and international technical affairs for the Motor Vehicle Manufacturers Association of the United States, told *Imponderables* that most manufacturers comply by using an unlock button, but others require the key to be pushed in toward the column to release it for withdrawal.

Submitted by William C. Nielsen of Rolling Hills Estate, California.

How Did Kodak Get Its Name? Is It True That the Name Comes from the Sound of the Shutter?

Standard Oil of New Jersey was roundly criticized when it changed its name to Exxon, a word chosen for its euphony and

DO PENGUINS HAVE KNEES?

uniqueness rather than any obvious meaning or associations. The name change was deplored by editorialists for confirming the soullessness of the post–World War II era.

But George Eastman trademarked "Kodak" in 1888. And the word didn't mean a darn thing.

The Eastman Kodak Company gets asked this question so often that they have prepared a pamphlet on the subject. Yes, they have heard the rumor that Kodak was the onomatopoeic description of the shutter closing, but the truth, although more prosaic, is fascinating nonetheless. As Eastman Kodak describes it, "George Eastman invented it out of thin air!" In several letters, Eastman described the less than scientific process that led to the creation of one of the most famous trademarks of the United States:

> I devised the name myself. . . . The letter "K" had been a favorite with me—it seems a strong, incisive sort of letter. . . . It became a question of trying out a great number of combinations of letters that made words starting and ending with "K." The word "Kodak" is the result.

> . . . it was a purely arbitrary combination of letters, not derived in whole or part from any existing word, arrived at after considerable search for a word that would answer all the requirements for a trademark name. . . . it must be short; incapable of being misspelled so as to destroy its identity; must have a vigorous and distinctive personality; and must meet the requirements of the various foreign trademark laws.

The last point is important. One can't receive a trademark if the proposed name is merely an existing word (or words) found in the dictionary that accurately describes the item. As Eastman wrote in 1906, "There is, you know, a commercial value in having a peculiar name."

Submitted by John M. Clark of Levittown, New York.

DAVID FELDMAN

Why Are There Ridges (Often Painted Black) on the Sides of Most School Buses?

Those ridges are called "rub rails," and their main purpose is to add strength to the sides of the vehicle. According to John Chuhran, of Mercedes-Benz of North America: "If they were flat, the sides of buses would have more of a tendency to flex."

Rub rails have an additional benefit. Pete James, regional director of the National Association for Pupil Transportation, told *Imponderables* that rub rails also help maintain the structural integrity of the bus when it is broadsided. The rub rails help prevent the crashing vehicle from penetrating farther into the bus.

Why are the rub rails painted black? No, they are not racing stripes. The answer is murky indeed. Although there is a national standard mandating that school buses be painted "national school bus chrome yellow," some states require that both the rub rails and bumpers be painted a contrasting color to "stand out" from the yellow. In most cases, this color is black. Different sources attribute the choice of black to everything from aesthetics to hiding marks on the school bus to differentiating the rub rails from the main panel of the bus to making the school bus look different than other buses.

Pete James feels that the black stripes actually detract from the instantaneously recognizable look of school bus yellow. Karen E. Finkel, executive director of the National School Transportation Association, notes that several states specify that their buses may not be painted black. Yet Kentucky specifies that "the area between the window rub rail and the seat rub rail be painted black so that the white letters for the school district name will show up better." In matters of color choice, states' rights still prevail.

Submitted by Jens P. Aarnaes of Schaumburg, Illinois.

Why Do All the Armed Forces Start Marching with the Left Foot? Is There Any Practical Reason? Is This Custom the Same All Over the World?

Bottom line: We can only answer the third question with any confidence. As far as we can ascertain, soldiers all over the world step off on the left foot.

We contacted many of our trusty military sources about why the practice spread. They collectively shrugged their shoulders.

Imponderables has been besieged by questions about the origins of left/right customs (e.g., why we drive on the right side of the road, why the hot water faucet is on the left, why military medals are worn on the left) and found that usually the practices stem from a technical advantage.

What possible advantage could there be in starting a march with the left foot? We received a fascinating speculation on the subject by Robert S. Robe, Jr., president of The Scipio Society of Naval and Military History, which may not be definitive but is certainly more sensible and interesting than anything else we've heard about the subject.

> When warfare was institutionalized in prebiblical times so that trained armies could fight one another on a battlefield, the evolution of infantry tactics in close formation required regimented marching in order to effectively move bodies of heavy and light infantry into contact with an enemy.
>
> I am hypothesizing that some long forgotten martinet discovered by accident or otherwise that a soldier advancing at close quarters into an enemy sword or spear line could, by stepping off on his foot in unison with his fellows, maintain better balance and sword contact to his front, assuming always that the thrusting or cutting weapon was wielded from the *right* hand and the shield from the left. The shield would also protect the left leg forward in close-quarter fighting.

Robe's explanation echoes the usual explanation of why we mount horses from the left. The horse itself couldn't care less

DAVID FELDMAN

from what side its rider mounts it. But in ancient times, when riders wore swords slung along the left side of the body (so that the swords could be unsheathed by the right hand), riders found it much easier to retain their groin if they mounted on the left.

Submitted by John Williams of North Hollywood, California. Thanks also to Ann Czompo of Cortland, New York, and Sylvia Antovino of Rochester, New York.

Why Do Rabbits Wiggle Their Noses All the Time?

Rabbits don't wiggle their noses *all* the time, but enough to make one wonder if they have a cocaine habit or a bad allergy. Little did we suspect that this charming idiosyncrasy is a key to the workings of the rabbit's respiratory system. There are at least four reasons why rabbits' noses twitch away.

1. The movement activates the sebaceous gland (located on the mucus membranes) and creates moisture to keep the membranes dampened and strong. Furthermore, like other animals (including, ahem, dogs with wet noses), rabbits can smell better off of a wet surface.

2. Frequent wiggling expands the nasal orifices, or nares, so that the rabbit can inhale more air. Like dogs with wet noses, domestic rabbits can't perspire. According to Dr. T. E. Reed, of the American Rabbit Breeders Association, nose wiggling actually helps rabbits cool themselves off on hot days:

> The only method of cooling themselves is by expiring the super heated air from the respiratory tract to the environment and through the convection of heat from the ears. . . . The inhalation . . . of a voluminous amount of air is extremely important.
>
> The normal respiratory rate in the domestic rabbit is approximately 120 breaths per minute. However, during extremely hot weather, it is not uncommon for the respiratory rate to approach 300 to 350 breaths per minute.

The nares [control] the amount of air rabbits can inhale. In order to increase the volume of air that is inhaled, the rabbit will twitch its nose by activating the various types of muscles surrounding each of the nostrils to increase the orifice size.

3. When a rabbit's whiskers are touched, the muscles surrounding the nostrils expand and contract in order to sharpen the animal's olfactory abilities.

4. If a rabbit continuously wiggling its nose appears to be nervous, T. E. Reed reminds us that it might be:

> When a rabbit is calm and unattended, each of the nostrils usually will remain in a stationary position. However, if the rabbit gets excited, the rabbit's pulse rate and respiratory rate increase and there is a nervous intervention to the nose that causes a constriction and relaxation of the paranasal muscle—the "wiggle" that most lay individuals observe.

Although some readers have speculated that the "wiggle" is caused by the continual growth of the incisor teeth, each of the rabbit experts we spoke to disputed the claim.

Submitted by Garnett Budd of Eldorado, Ontario. Thanks also to Jason Gonzales of Albuquerque, New Mexico.

Why Are There Legless Ducks in the Crest of Cadillacs?

Jil McIntosh, who writes for several automotive magazines and owns a 1947 Cadillac, told us that two rumors abound about the ducks on the Cadillac crest: that the six ducks signify the original founders of the Cadillac company; and that each duck stands for one of the six cylinders.

These rumors are wrong on all counts. Cadillac was founded by one person, a Civil War riflemaker, Henry Leland, in 1902. After selling the company to General Motors, Leland proceeded to introduce the Lincoln, which later became Cadillac's main luxury rival.

McIntosh laughs off the cylinder theory:

> The first Cadillacs were one-cylinder, and then four-cylinder. In 1915, Cadillac brought out their first V-8 engines. . . . No six-cylinder engine was ever used until the Cimarron of the mid-1980s (and I hope some Caddy stylist rots in hell for *that* one!). So the cylinder theory is out; and since Leland was the main brain, the partnership theory is out. The crest is actually a coat of arms.

But not just anybody's coat of arms. The Cadillac is named after Antoine de la Mothe Cadillac, the French explorer who founded the city of Detroit in 1701. McIntosh told *Imponderables* that "the Cadillac coat of arms crest is a version of Cadillac's family crest, which had been in his family for 400 years prior to his birth."

Those aren't ducks in the crest—they are merlettes, heraldic adaptations of the martin (a kind of swallow). The merlettes on the Cadillac crest have no beaks or legs. When merlettes appear in threes, they refer to the Holy Trinity.

The Cadillac division receives many queries about the birds, and has even explained why the merlettes are missing legs:

> The merlettes were granted to knights by the ancient School of Heralds, together with the "fess" [the wide, dark horizontal band separating the two upper birds from the one on the bottom of the quarterling], for valiant conduct in the Crusades. The birds shown in black against a gold background in this section of the Cadillac arms denote wisdom, riches, and cleverness of mind, ideal qualities for the adventurous and zealous Christian knight.
>
> Of the merlette, Guillaume, an ancient historian, says: "This bird is given for a difference, to younger brothers to put them in mind that in order to raise themselves they are to look to the wings of virtue and merit, and not to the legs, having but little land to set their feet on."

The Cadillac crest has graced every Cadillac ever built,

often in several places. The crest has changed cosmetically from time to time—it has been lengthened or thinned, adorned with laurel leaves, etc. But the heraldic design has remained the same.

We hope we have answered the Imponderable once posed by Chico Marx: Viaduct(s)?

What Is the Technical Definition of a Sunset or Sunrise? How Is It Determined at What Time the Sun Sets or Rises? Why Is There Natural Light Before Sunrise and After Sunset?

The definitions are easy. A sunrise is defined as occurring when the top of the sun appears on a sea-level horizon. A sunset occurs when the top of the sun goes just below the sea-level horizon.

But how do scientists determine the times? No, they do not send meteorologists out on a ladder and have them crane their necks. No observation is involved at all—just math. By crunching the numbers based on the orbit of the Earth around the Sun, the sunrise and sunset times can be calculated long in advance.

Richard Williams, a meteorologist at the National Weather Service, explains that published times are only approximations of what we observe with our naked eyes:

> The time of sunrise and sunset varies with day of the year, latitude, and longitude. The published sunrise and sunset times are calculated without regard to surrounding terrain. That is, all computations are made for a sea-level horizon, even in mountainous areas. Thus the actual time of sunrise at a particular location may vary considerably from the "official" times.
>
> When we observe sunset, the Sun has already gone below the horizon. The Earth's atmosphere "bends" the Sun's rays and delays the sunset by about three minutes. Likewise with sunrise, the sun makes its first appearance before it would on a planet with no

DAVID FELDMAN

atmosphere. We actually get five to ten minutes of extra sunlight due to this effect.

Submitted by a caller on the Larry Mantle show, Pasadena, California.

Why Is Pubic Hair Curly?

In *Why Do Clocks Run Clockwise?*, we discussed why we have pubic hair. But you weren't satisfied. So we will continue our nonstop exploration of what seems to be an insatiable North American interest in body hair.

If you want to know the anatomical reason why pubic hair is curly, we can help you. Dr. Joseph P. Bark, diplomate of the American Board of Dermatology, explains:

> Pubic hair is curly because it is genetically made in a flat shape rather than in a round shape. Perfectly round hair, such as the hair seen on the scalps of Native Americans, is straight and has no tendency to curl. However, ribbonlike hair on the scalps of blacks is clearly seen to curl because it is oval in construction. The same is true with pubic hair. . . .

But answering what function curly pubic hair serves is a much trickier proposition. Some, such as Samuel T. Selden, a Chesapeake, Virginia, dermatologist, speculate that pubic hair might be curly because if it grew out straight and stiff, it might rub against adjacent areas and cause discomfort. (Dermatologist Jerome Z. Litt, of Pepper Pike, Ohio, who has been confronted with the question of why pubic and axillary hair doesn't grow as fast as scalp hair, facetiously suggests that "not only wouldn't it look sporty in the shower room, but we'd all be tripping over it.")

Before we get carried away with our theories, though, we might keep in mind a salient fact—not all pubic hair is curly.

DO PENGUINS HAVE KNEES?   177

Early in puberty, it is soft and straight. And Selden points out that if this book were published in Japan or China, this Imponderable likely would never have been posed. The pubic hair of Orientals tends to be sparser and much straighter than that of whites or blacks.

Submitted by Suzanne Saldi of West Berlin, New Jersey.

Why Are There Tiny Holes in the Ceiling of My Car?

For the same reason there are tiny holes in the ceiling of many schools and offices. They help kill noise. Chrysler's C. R. Cheney explains:

> The headliners in some automobiles and trucks have small perforations in them to help improve their sound-absorbing qualities. The perforated surface of the headliner is usually a vinyl or hardboard material and it is applied over a layer of foam. The holes serve to admit sound from inside the vehicle and allow it to be damped by the foam layer to promote a quieter environment for passengers.
>
> To some, the patterns made by the tiny perforations were also pleasing to the eye, so perhaps the perforations served double duty.

Let's not stretch it, C.R.

Submitted by Garland Lyn of Windsor, Connecticut.

DO PENGUINS HAVE KNEES?

What Does the "YKK" Emblazoned on My Zipper Mean?

It means that you are the proud possessor of a zipper made by YKK Inc. (That *is* why you bought that pair of pants, isn't it?) Now perhaps YKK's emblem is a little less picturesque than an alligator or a polo player, but then, who except *Imponderables* readers busy themselves by reading their zippers, anyway? And the subdued logo hasn't seem to hurt YKK's business; Izod and Ralph Lauren would kill for the 75+ percent share of their market that YKK commands.

YKK, the largest zipper manufacturer in the world, stands for Yoshida Kogyo Kabushikikaisha (now you know why they call themselves YKK). Yoshida is the last name of the founder of the company, Tadao Yoshida. In English, YKK is translated as Yoshida Industries Inc.

Submitted by Juli Haugen of West Boothbay Harbor, Maine.
Thanks also to Gwen Shen of San Francisco, California;
M. Sullivan of Miami, Florida; Becky Wrenn of Palo Alto,
California; Tisha Land of South Portland, Maine; Anne
Daubendiek of Rochester, New York; Chris Engeland of Ottawa,
Ontario; and many others.

Why Do VCR Manuals Advise You to Disconnect the Machine During Storms?

If lightning strikes your home and your antenna or AC power line is not properly grounded, you can find yourself with a busted VCR. William J. Goffi, of the Maxell Corporation, explains:

> ... the electrical surge will find its way through your home's wiring and into your VCR, and consequently cause much damage to your unit. This is true, however, for any electrical appliance you

have. A surge of lightning can blow out your television monitor, your stereo, etc.

Purchasing a surge protector will protect your investments.

As Mark Johnson, a spokesperson for Panasonic, points out, an electrical surge can blow out an electrical appliance even if it is off, which is why the manual recommends disconnecting the plug.

Another potential threat to a connected appliance is a sudden surge of power from the local electric power utility. Surge protectors will also usually solve this problem.

Submitted by Richie, a caller on the Dan Rodricks show, Baltimore, Maryland.

In Baseball, Why Is the Pitcher's Mound Located 60′6″ from Home Plate?

The answer comes from Bill Deane, senior research associate of the National Baseball Hall of Fame:

> The pitcher's box was originally positioned 45 feet from home plate. It was moved back to 50 feet in 1881. After overhand pitching was legalized, it was moved back to 60′6″ in 1893.

Why was the mound moved back? For the same reason that fences are moved in—teams were not generating enough offense. Morris Eckhouse, executive director of the Society for American Baseball Research, told *Imponderables* that around the turn of the century, batters were having a hard time making contact with the ball.

But what cosmic inspiration led to the choice of 60′6″ as the proper distance? Deane says there is evidence indicating that "the unusual distance resulted from a misread architectural drawing that specified 60′0″."

Submitted by Kathy Cripe of South Bend, Indiana.

Why Does Grease Turn White When It Cools?

You finish frying some chicken. You reach for the used coffee can to discard the hot oil. You open the lid of the coffee can and the congealed grease is thick, not thin, and not the yellowish-gold color of the frying oil you put in before, but whitish, the color of glazed doughnut frosting. Why is the fat more transparent when it is an oil than when it is grease?

When the oil cools, it changes its physical state, just as transparent water changes into more opaque ice when it freezes. Bill DeBuvitz, a longtime *Imponderables* reader and, more to the point, an associate professor of physics at Middlesex County College in New Jersey, explains:

> When the grease cools, it changes from a liquid to a solid. Because of its molecular structure, it cannot quite form a crystalline structure. Instead, it forms "amorphous regions" and "partial crystals." These irregular areas scatter white light and make the grease appear cloudy.
>
> If grease were to solidify into a pure crystal, it would be much clearer, maybe like glass. Incidentally, paraffins like candle wax behave just like grease: They are clear in the liquid form and cloudy in the solid form.

Submitted by Eric Schmidt of Fairview Park, Ohio.

Why Is the Skin Around Our Finger Knuckles Wrinkled When the Skin Covering Our Knees Is Not?

We received this Imponderable about three years ago, in a stack of letters from Judith Bambenek's South St. Paul High School class. No doubt, her students were bludgeoned into writing us, but we were nevertheless impressed by the quality of the ques-

DAVID FELDMAN

tions. By now, we're sure that Chris Dahlke is on the way to becoming a Rhodes Scholar.

Dr. Harry Arnold, Jr., a distinguished dermatologist from the land of Rice-A-Roni, was happy to solve the Imponderable troubling the youth of South St. Paul:

> In extension, the knuckles need enough skin to permit flexing the joint roughly through 100 degrees, so there is excess skin when the joint is fully extended.
>
> The knees require much less skin but there is wrinkling there too, over a much larger area, so it is less obvious. Even with the extra skin, we get "white-knuckled" when the joints in the knee are fully flexed.

We're sometimes amazed at the lengths to which our sources will extend themselves for the sake of science and the vanquishing of Imponderability. In the case of Chesapeake, Virginia, dermatologist Samuel T. Selden, it included disrobing. Selden has a speculative but fascinating anthropological theory to explain the knee-finger disparity:

> I had to take off my shoes and socks to check, but interestingly, the skin over the knuckles of our toes is not very wrinkled either. The skin over our elbows is corrugated, but not to the degree that the skin over the finger knuckles is wrinkled.
>
> My theory for the wrinkling is that our ancestors, the apes, walked on their fingers, as we probably did prior to becoming upright beasts. The wrinkles are most apparent over the middle knuckles, the proximal interphalangeal joints, where apes place most of their weight when walking. Some individuals, through heredity, have thickened skin in this area known as "knuckle pads," and they are probably even more of a throwback to their ape ancestors.

We're sure they'll be thrilled to hear that.

Submitted by Chris Dahlke of South St. Paul, Minnesota. Special thanks to Chris's teacher, Judith Bambenek.

DO PENGUINS HAVE KNEES? 183

Where Does All of the Old Extra Oil in Your Car's Engine Lurk After an Oil Change?

Our befuddled correspondent, Victor Berman, elaborates:

> Just before you change your oil you can check the dipstick. The oil level is "full." You then drain the oil and change the filter, put in the recommended amount of oil, and check the dipstick. The level is "full."
>
> Now you go to dispose of the old oil from the crank case and the filter and, lo and behold, there is less than five quarts. More like three to three and one-half quarts. I know that even after turning over the filter and letting it drain there is some oil left in the filter, but not one and one-half quarts. Is my car's engine storing an extra quart and a half every time I change the oil?

We were intrigued with this mystery, so we contacted several auto manufacturers, who had no explanation for the case of the missing oil. So we persisted, engaging in two long conversations with oil specialists: H. Dale Millay, a research engineer for Shell Oil, and Dan Arcy, a technical service representative for Pennzoil Products Company. After much soul-searching, all of us decided we still had an Imponderable, bordering on a Frustable, on our hands.

Some questions don't yield one simple answer. So the experts ventured several possible explanations:

> 1. If the oil change is conducted while the engine is cold, the oil will be thicker and tend to sit on the motor's surface and coat internal surfaces. Even hot oil will wet the internal surfaces and result in some oil loss.
>
> 2. The amount of oil unleashed depends to a great extent upon the location of the plug on the drain. Dan Arcy points out that Ford, for example, manufactures several models with two drain plugs—one needs to pull both plugs to get rid of all the oil.
>
> 3. In some cases, the slant of the car may inhibit or promote freer flow of oil out of the drain. Any flat reservoir has to be tipped over to spill out all of the contents.

DAVID FELDMAN

4. Are you sure you drained the oil filter adequately? Millay thinks the oil filter, which is built to hold up to two quarts of oil, is the most likely hiding place for most of the missing oil.

5. Oil will continue to drizzle out of the plug a long time, often an hour or more. This doesn't explain the loss of a quart and one-half, but then every drip counts when trying to solve this Imponderable.

6. A significant amount of oil may be left on your oil pan. Not a quart, perhaps, but a half-pint or so may be underestimated if spread around a pan with a large circumference.

7. Not to challenge your dipstick-reading acuity, Victor, but our experts wanted to ask you if you are sure you were really checking the oil directly before changing it. All engines are designed to consume some oil when operating.

8. How about a mundane reason? A leak? Arcy relayed an astonishing maxim of the industry: The loss of one drop of oil every fifty-five feet is equivalent to the consumption of one quart of oil in 500 miles. Of course, the leak theory doesn't explain why the oil shortfall occurs only when changing the oil.

Any readers have a solution to this greasy Imponderable?

Submitted by Victor Berman of East Hartford, Connecticut.

HE FLOATS SO NATURAL-!

Why Do Fish Float Upside-Down When They Die?

Imponderables cannot be held responsible for the consequences if you read this answer within thirty minutes of starting or finishing a meal. With this proviso, we yield the floor to Doug Olander, director of special projects for the International Game Fish Association:

> Fish float upside-down when they die because internal decomposition releases gases that collect in the gut cavity. Anyone who's ever cleaned a fish knows the meat is on top (dorsally) and the thin stomach wall on the bottom (ventrally). So as gases accumulate, the dense muscle mass of the top of the fish is positioned down and the gas-filled stomach up.
>
> Fishes with swim bladders already have gas inside, which tends to make them at least neutrally buoyant. Benthic fishes, lacking swim bladders (flatfishes, for example) would *not* float upon death.

　　　　　　　　　　　　　DAVID FELDMAN

Deepwater fishes float high atop the surface when pulled rapidly upward, a common angling experience, because the gas trapped inside their swim bladder expands at the reduced pressure of the surface.

Dr. Robert Rofen, of the Aquatic Research Institute, adds that since so much of a fish's body weight is concentrated along the bone structure of the back and skull, it is not uncommon to find dead fish floating with heads down.

Submitted by Melissa Hall of Bartlett, Illinois.

Why Do Some Companies Use Mail-In Refunds Rather Than Coupons?

Applying our usual paranoid logic, we always assumed that more people will redeem coupons at a grocery store than will bother tearing off proofs of purchase and mailing in forms to receive a refund. Therefore, a mail-in refund's purpose in life was to seduce you into buying eight cans of tuna but then being too lazy to ever send in the proofs of purchase and cash register receipt to receive the rebate.

We remember once soaking pineapple cans in hot water, trying to peel labels off to send as proofs of purchase, and wondering: "Is this why we were put on earth? There must be a better way." But there is logic in marketers' refund nonsense.

F. Kent Mitchel, chairman of the Marketing Science Institute, confirmed our conspiracy theory:

> Mail-in refunds are generally less expensive largely because of lower usage by the public, yet they protect existing brands about as well as coupons in a competitive situation.

What does Mitchel mean by "protect"? In many cases, coupons are used to promote items that consumers consider as commodities, with insubstantial differences in quality, and where brand loyalty may not withstand a pricing differential. Pepsico and Coca-Cola wage perpetual price wars in the stores and through coupons. A similar skirmish invades the detergent and coffee aisles. Coupons and mail-in refunds, then, are often used to "protect" one brand against price cuts by competing brands.

In many cases, the cash reward for mail-in refunds is higher than those for coupons, but the lower redemption rates make mail-ins cheaper in the long run. As Robert A. Grayson, publisher of *The Journal of Consumer Marketing,* told *Imponderables,* "the promotion looks as big but doesn't cost as much," particularly if consumers purchase the goods and neglect to ever send for the rebate.

But cost isn't the primary consideration in implementing a mail-in rather than a coupon campaign. The choice is really a strategic decision dictated by whom the marketer is trying to attract. Thomas L. Ruble, consumer response manager of the Louis Rich Company, explains:

> Coupons are used to stimulate new business—to encourage first-time buyers. Mail-in refunds, on the other hand, encourage continuity among the established customer base. Mail-ins also encourage established customers to purchase multiple packages.

Mail-in refunds are also most effective for products, including foods, sold outside of grocery stores. Supermarkets are geared for the paperwork involved in processing coupons. But a family-run hardware or camera store might not know how to receive compensation for the refund on a package of batteries or be willing to put up with the nuisance of doing so.

One other crucial point. By making you fill out personal information for the refund, the marketer now has in its possession your name and address. Most companies retain this information in databases, and then can ply you with direct-mail campaigns.

Why Do Grocery Coupons Have Expiration Dates?

Why are some grocery coupons effective for only a few weeks? Why would the marketer spend so much money, not only in redeeming coupons but in placing them in newspapers, only to invalidate the coupons so quickly?

Usually, the expiration date is added for the same reason a deadline was placed on when your term paper in school was due: Marketers, like teachers, know that you need a cattle prod and the threat of a deadly weapon to motivate you to act in the "right" way.

Occasionally, a company might want to spur sales, either because the brand is in danger of losing its shelf space if sales don't improve or because the company (or a particular executive) needs to demonstrate sales growth in a short period. As we have already said, coupons can be used as a preemptive price-cutting strike against new (or old but gaining) competition.

But another, more important financial consideration plagues food marketers, one that threatened the financial stability of the airlines after frequent flier programs were instituted. F. Kent Mitchel, chairman of the Marketing Science Institute, explains:

> Expiration dates reduce the liability of the float. Only a small percentage of coupons are ever redeemed and a company budgets to cover the expected redemption. There are literally billions of coupons floating around at any time and if all the coupons that a company issued were redeemed, it would be an enormous unanticipated expense and could quite possibly bankrupt even a large company.
>
> To avoid this unsavory possibility, coupons are rendered valueless after a certain time by using expiration dates. It has been my experience that most major manufacturers will redeem coupons beyond their expiration dates if presented with the proper proof of purchase.

Maybe. But will the supermarkets redeem them?

Submitted by Linda Harris of Holbrook, Maine.

Why Do Only Female Mosquitoes Eat Human Blood? What Do Male Mosquitoes Eat?

No, the mosquito menfolk aren't out eating steak and potatoes. Actually, the main food of both male and female mosquitoes is nectar from flowers. The nectar is converted to glycogen, a fuel potent enough to provide their muscles with energy to fly within minutes of consuming the nectar. Mosquitoes also possess an organ, known as the fat-body, that is capable of storing sugar for conversion to flight fuel.

Male mosquitoes can exist quite happily on a diet of only nectar, and nature makes certain that they are content—males don't have a biting mouth part capable of piercing the skin of a human. But females have been anatomically equipped to bite because they have an important job to do: lay eggs. In some species, female mosquitoes are not capable of laying any eggs unless they eat a nutritional supplement of some tasty, fresh blood. Their organs convert the lipids in blood into iron and protein that can greatly increase their fecundity.

A mosquito that would lay five or ten eggs without the supplement can lay as many as 200 with a dash of Type O. Although we don't miss the blood sucked out of us, this is quite a feast for the mosquito; many times, she consumes more than her own body weight in blood.

But let's not take it personally. Some studies have indicated that given a choice, mosquitoes prefer the blood of cows to humans, and in the jungle are just as likely to try to bite a monkey or a bird as a human.

Submitted by Carolyn Imbert of Yuba City, California.

If the Third Prong on an Electrical Plug Is for Grounding and Shock Protection, Why Don't All Plugs Have Three Prongs?

In the good old days, electrical plugs had two prongs and the receptacles were ungrounded. If you happened to use the wrong side of the circuit, it could be a shocking experience. So a simple and effective idea was developed: add a third prong. Don French, chief engineer for Radio Shack, explains the principle:

> If any short circuit developed in the wiring or device being powered, then instead of shocking the next person who touched the device, the third prong, being grounded, would carry the current until a fuse would blow. Now it is common to find three-pronged plugs on most portable and stationary appliances.

Meanwhile, however, other engineers were working on "double insulated" prongs that do not require a third prong for protection. Although the third prong was rendered unnecessary, most old receptacles weren't wide enough to receive the fatter prongs—another example of incompatible technologies that benefited the manufacturers (think of all the consumers who had to refit receptacles and buy new extension cords to hold double insulated prongs) and bankrupted the consumer.

Submitted by Ronald C. Semone of Washington, D.C. Thanks also to Terry L. Stibal of Belleville, Illinois; David A. Kroffe of Los Alamitos, California; Margaret K. Schwallie of Kalamazoo, Michigan; Kurt Dershem of Holland, Michigan; Layton Taylor of Yankton, Michigan; and Robert King of Newbury Park, California.

Why Does Menthol Feel Cool to the Taste and Cool to the Skin?

Of course, the temperature of menthol shaving cream isn't any lower than that of musk shaving cream. So clearly, something funny is going on. R. J. Reynolds's public relations representative, Mary Ann Usrey, explains the physiological shenanigans:

> The interior of the mouth contains many thermoreceptors that respond to cooling. These thermoreceptors may be compared to the receptors for the sensations of "sweet," "salty," "bitter," etc.
>
> In other words, individual receptors respond to specific types of stimulation. For example, a person's perception that sugar is sweet is initiated when the receptors in the mouth for "sweet" are stimulated. Menthol feels cools to the taste because menthol stimulates the thermoreceptors that respond normally to cooling.
>
> Menthol has the ability to "trick" those thermoreceptors into responding. The brain receives the message that what is being experienced is "cool."
>
> Although not as easy to stimulate by menthol as those in the mouth, the skin also contains those types of thermoreceptors, which is why menthol shaving cream or shaving lotion feels cool to the skin.

Submitted by Allan J. Wilke of Cedar Rapids, Iowa.

Why Do Bridges Freeze Before Nearby Roads?

Asphalt, used in most roads, retains heat better than bridge decks, which are usually made out of concrete slabs. But the most important reason has more to do with elementary physics. Stanley Gordon, chief of the Bridge Division of the Federal Highway Administration, explains:

> A bridge deck will freeze before a roadway pavement because it is exposed to the environment from both the top and the bottom sides. In contrast, a roadway pavement is only exposed to the environment from the top side.

In other words, the earth itself provides insulation to roads. Any heat that accumulates in the bridge during the day will be released as the ambient temperature drops. As Amy Steiner, program director for the American Association of State Highway and

DO PENGUINS HAVE KNEES?

Transportation Officials, put it, "Bridge decks can release only heat absorbed by the deck itself and obviously do not benefit from the heat retained by the soil."

Submitted by Roger Mullis of Eureka, California.

Chew

... *it won't lose its flavor you-know-where!*

Why Can't They Make the Flavor in Chewing Gum Last Longer?

Call us paranoid, but we always suspected that gum manufacturers attended trade seminars on such subjects as "The Enemy: Long-Lasting Flavor," "How to Make Sure Your Customers' Chewing Gum Loses Its Flavor on the Bedpost Overnight," and "How to Make Your Gum Tasteless Before Your Sucker Customer Has Thrown Away the Wrapper." Emboldened by such rhetoric, the gum makers see dollar signs floating above their eyes and produce gum whose flavor lasts less time than a Zsa Zsa Gabor marriage. Naive consumers are then confronted with the imperative of plopping another stick of gum into their mouths to receive the flavor jolt they received all of, maybe, three minutes ago.

But industry folks insist that the conspiracy theory just isn't true. In fact, Bill O'Connor, director of administration at the Topps Company, told *Imponderables* that if a company could create a gum that retained flavor longer, it would hammer this competitive advantage home in advertisements. Wrigley's has done just that with its Extra gum, which uses "encapsulation," little flavor pockets that require more mouth action than conventional gum to draw out its flavor. In essence, Wrigley created a time-release gum.

But even encapsulation doesn't beat the two main enemies of flavor retention in chewing gum:

1. The saliva generated from chewing literally drags the flavoring (and sugar) out of the gum.
2. The mouth gets fatigued and sensitized to any flavor eventually.

O'Connor suggests that if you put aside a "used" piece of gum, eat a saltine to cleanse the palate, and then plop the gum back into your mouth, it will taste flavorful again.

May we suggest an alternative: plopping a new stick of gum in your mouth.

What Is the Purpose of the Plastic Bags in Airline Oxygen Masks When They Don't Inflate?

We're always amazed when we find out that an airplane has been evacuated successfully during an emergency landing. The airlines try to do a good job briefing passengers on the safety requirements before takeoff. But a quick scan of the passengers will indicate that the seasoned fliers are already napping or deeply engrossed in the scintillating inflight magazine, while the less experienced tend to be hanging on every word, in a panic, trying to conjure in their minds how they can convert their seat cushion into a flotation device.

We tend to combine the worst aspects of both types of passengers. We attempt to read our newspaper, having heard the announcement 80 million times, but we're actually trying to sup-

DAVID FELDMAN

press our fear that there aren't *really* oxygen masks up there that are going to drop down during an emergency.

All white-knucklers are familiar with the proviso in the safety demonstrations of oxygen masks: "Although the bag won't inflate, oxygen is still flowing . . ." or the variant, "Although the bag will not *fully* inflate . . ." Several sharp *Imponderables* readers have wondered: If the bag doesn't inflate, why does it have to be there? Our image of an oxygen bag comes from *Ben Casey*, where resuscitators inflated, deflated, and reflated as violently as a fad dieter.

But the bag does serve a purpose. Honest. The mask used by airlines is called a "phased-dilution" mask. As you inhale, you are breathing in a mixture of ambient air and oxygen. Compressed oxygen is quite expensive, and particularly at low altitudes, you actually need very little pure oxygen even if the cabin is depressurized.

A nasty little secret is that a bizarre cost-saving device, the "oxygen mask" used in safety demonstrations, is not the real thing (if you look carefully, on most airlines, the mask will be marked "DEMO") and isn't even an exact replica. The real oxygen mask contains three valves that are the key to regulating your breathing in an emergency. The first, interior, valve pumps in pure oxygen. When the oxygen is depleted, the valve closes and the second, exterior valve opens and brings in ambient air (thus the term "phased-dilution"). The third, external valve, with a spring device, opens only to allow you to vent your exhalation.

According to oxygen equipment expert David DiPasquale, an engineer and administrative and technical consultant and major domo at DiPasquale & Associates, the normal cabin pressure is set to simulate the atmosphere of approximately 8,000 feet. The oxygen system automatically adjusts to different altitudes, varying the flow of oxygen. The higher the altitude, the higher percentage of oxygen (to ambient air) and the faster the flow rate of oxygen is required. During decompression, a plane may suddenly find itself at an atmosphere equivalent to the ambient air at 35,000 feet or higher.

DO PENGUINS HAVE KNEES?

The bottom line is that there is no reason on earth why the plastic bag should inflate dramatically. The oxygen bag itself might hold about a liter and one-half of gas. At 18,000 feet, the system might pump in about one liter per minute; at 40,000 feet, about three liters per minute. But unlike the *Ben Casey* resuscitator, only a small percentage of this gas is inhaled in any one breath.

At higher altitudes, the bag will noticeably inflate, both because the flow rate of oxygen is much higher and because the bag has a natural tendency to expand when air pressure is lower. As Richard E. Livingston, of the Airline Passengers Association of North America, put it:

> Since oxygen, like other gases, expands at higher altitudes, maximum inflation will be obvious at high altitude. Conversely, gases are more compressed at low altitudes, so little or no bag inflation will be evident at lower altitudes.

Submitted by Charles Myers of Ronkonkoma, New York. Thanks also to Mick Luce of Portland, Oregon, and Stanley Fenvessy of New York, New York. Special thanks to Jim Cannon of Lenexa, Kansas.

DAVID FELDMAN

Frustables

The 10 Most Wanted OR Imponderables

We don't claim to be infallible. In fact, most of the time, we are experts in fallibility. Too often for our satisfaction, readers confront us with fascinating Imponderables that we cannot answer. These Imponderables tend to fall into two groups: mysteries that totally baffle the experts we consult; or mysteries that every expert has an opinion on, but for which there is no consensus.

Either way, our inability to answer these questions makes us frustrated. So, we throw Frustables (short for "frustrating Imponderables") out to you in the fervent hope that you can do better than we can. It's amazing how often you can, as demonstrated in the Frustables Update section that follows.

To lure you into sharing your wisdom, we offer a complimentary, autographed copy of our next volume of *Imponderables* to the first person who can lead to the proof that solves any of these Frustables. And of course, your contribution will be displayed and acknowledged in the book.

But don't get smug until you see the new Frustables. Solving these will not be easy.

FRUSTABLE 1: *Why Do Doctors Have Bad Penmanship?*

Even physicians we contacted agreed that the stereotype is, more often than not, true. You wouldn't believe how many theories we've heard to explain/deplore/rationalize/excuse this phenomenon. We have been able to confirm that no medical school in the United States offers a specific course on bad penmanship. So is there any other explanation?

FRUSTABLE 2: *Why Are Salt and Pepper the Standard Condiments on Home and Restaurant Tables? When and Where Did This Custom Start?*

We look upon salt and pepper on the table as being as inevitable as the plate and silverware. But it didn't have to be that way.

FRUSTABLE 3: *Why Don't People Wear Hats as Much as They Used to?*

The comeback of the hat has been bandied about as much as the return of big bands. But it never seems to happen. We have millions of theories about this but no consensus has emerged. Have any *Imponderables* readers given up wearing hats? If so, why?

FRUSTABLE 4: *How and Why Were the Letters B-I-N-G-O Selected for the Game of the Same Name?*

Before bingo, many similar games existed with different names.

FRUSTABLE 5: *Why Do They Always Play Dixieland Music at Political Rallies When Dixieland Isn't Particularly Burning Up the Hit Parade at the Moment?*

Do political consultants hire Dixieland bands because that's what politicians have always done? Is Dixieland the least objectionable musical form? If so, why don't you hear it more often on the radio?

FRUSTABLE 6: *Why Does Eating Ice Cream Make You Thirsty?*

Most of the taste experts and ice cream makers we've contacted deny that the premise of this question is true. But we've received the question several times and experienced the sensation ourselves. We even had a friend who loved malts and would drink one and then order an iced tea to quench his thirst.

FRUSTABLE 7: *Why Are Belly Dancers So* Zaftig?

By Western standards, belly dancers are rather fleshy around the midriff, surprising in artists who are constantly exercising this region. Experts we've contacted differ violently on this subject.

DAVID FELDMAN

Some say that the muscles contracted to belly dance are not those that would make the belly look Sheena Eastonish. Some say that standards of beauty in the Middle East are different and that most dancers deliberately keep some flesh. And others denied the premise. What do you think?

FRUSTABLE 8: *How Was Hail Measured Before Golf Balls Were Invented?*

Okay, we admit we're being facetious here, but we would be interested if any readers have heard hail compared to *anything* besides a ball (golf balls and baseballs are about all we ever hear) by local weathercasters.

FRUSTABLE 9: *Why Did 1930s and 1940s Movie Actors Talk So Much Faster Than They Do Today?*

Compare a Katharine Hepburn–Cary Grant comedy or a Bogie–Bacall melodrama with their contemporary counterparts, and they sound like a 45-rpm record playing at 78. What accounts for the huge change? We've heard tons of theories about this Frustable, too. But what are yours?

FRUSTABLE 10: *Why Does Meat Loaf Taste the Same in All Institutions?*

We admit that this is a personal obsession of ours. Ever since we noticed that meat loaf tasted the same in every school we ever attended, we've sampled the meat loaf any time we've been forced to eat at a cafeteria in an institution such as a federal building, hospital, or college. Does the government circulate a special *Marquis de Sade Cookbook?* Not all meat loaf tastes the same, but somehow the meat loaf at an elementary school in Los Angeles tastes the same as the meat loaf at a courtroom cafeteria in New York. Why does it?

Frustables Update

MERRY CHRISTMAS! HERE'S A FRUITCAKE! I JUST HATE IT, AND I DIDN'T WANT IT SITTING AROUND THE HOUSE!!

WHY, THANK YOU! I ALSO HATE IT! BUT, I CAN PALM IT OFF AS A GIFT TO SOMEONE ELSE, JUST LIKE YOU DID TO ME! NOW, WHO GAVE IT TO YOU?!

FRUSTABLE 1: *Does Anyone Really Like Fruitcake?*

As expected, we received more mail about fruitcake than all the other Frustables in *Why Do Dogs Have Wet Noses?* combined. When we posed this Frustable, we suspected that there wasn't one definitive answer to explain such a complex phenomenon as the perpetuation of this foodstuff, especially as a gift, that nobody seems to like. We were right.

Reader Bill Gerk, of Burlingame, California, was kind enough to point out that one of our favorite writers, Calvin Trillin, devoted a whole magazine column to this subject. Trillin claimed that "nobody in the history of the United States has ever bought a fruitcake for himself." Trillin was besieged with letters from readers claiming they had bought fruitcakes, "although the receipts are never enclosed."

Like Trillin, we can't offer proofs of purchase, but we certainly heard from fruitcake lovers. Scores of readers, including Lilet Quijano of Livermore, California, Edmund DeWan of Urbana, Illinois, Anne Wingate of Salt Lake City, Utah, and Betty Begley of Cambria, California, offered to accept the unwanted fruitcakes of *Imponderables* readers. We'd include the full addresses of these folks, but fear lawsuits if the offers were simply a sick joke.

Several fruitcake-loving readers tried to ingratiate themselves by claiming that the silent majority would grow to love fruitcake *if they only tried a good one*. Claire Manning of Brooklyn, New York, not only admits to liking fruitcake ("a noble, beloved, memory-evoking little piece of heaven") but to perpetrating said dessert on innocent friends and family:

> Consider yourself among the underprivileged for this omission in your poor life. I not only adore fruitcake but I *make* it every year at the winter holiday season and do occasionally give it as a gift. So far, I haven't *received* any . . . are they trying to tell me something?

Fruitcake can bring people together. Robert Tanner, of Naples, Florida, reports that he and his wife both love fruitcake. We are genuinely happy that they found each other but we must raise a sobering question: Should such a couple have children? Is the preference for fruitcake a hereditary trait? From the evidence of our mailbag, we think so.

Dorothy Lant, of Concord, New Hampshire, reports that her entire family likes fruitcake. Bisbee, Arizona's Judy R. Reis notes that her daughters do, her sisters do, and her parents do. Her son doesn't, but he only likes things with ketchup on them.

But fruitcake worship can cause family problems. Kim Anderson, of Alma, Arkansas, reports that because "My mom, my sister, my grandma, and my aunt like it, we always have fruitcake at Christmas, much to the dismay of me and the rest of the family members." But Kim's suffering is nothing compared to the shame of Melanie Morton, of Branford, Connecticut:

> Yes, there are people who like fruitcake. I believe this is an indication of mental imbalance. I offer as an example my father. He is overly fond of the stuff. In fact, he is not content to wait to be gifted with it. He actually goes out *in search of fruitcake!* As if

this is not enough, he hates chocolate. He's not allergic, mind you, he merely detests this wonderful creation.

Yes, fruitcake can wrench families apart.

So if the love of fruitcake is an unnatural preference, who is conspiring to foist this Milli Vanilli of foods upon us? Fred Steinberg, of Newton, New Jersey, thinks it is the evil of free enterprise. Fred once had a business professor who told the story of the marketing of the electric knife. Market research indicated from the start that consumers wouldn't actually use the appliance once they owned it. Still, they proceeded with the introduction of the product because it was a perfect present,

> a present for "kids" to give to their mom for Mother's Day, for people to give as shower gifts. An electric knife is not inexpensive, not expensive, and appears to be useful. That's why they were manufactured, bought by consumers, and now lie dormant in some remote drawer . . .

In other words, Steinberg's theory is that enterprising bakers have created a food designed to be given away rather than eaten. When you think of it this way, fruitcake is the ultimate diet food, since it is never actually consumed.

Some readers thought that fruitcake was a foreign conspiracy, with the English cited as the usual culprits. Given their reputation for fine cuisine, we are inclined to believe that the English invented fruitcake. Jennifer Beres, of Norwalk, Connecticut, actually sent us a sample of her mother's homemade fruitcake, which, even in the spirit of scientific investigation, we did not have the fortitude to sample. Jennifer veritably gushes with praise for the English art:

> My mother is British born and professes the fruitcake's existence originates in English tradition. Perhaps the reason that no one likes fruitcake is that the creation of a fruitcake is an art not to be duplicated in commercial factories by swiftly moving assembly lines. In order for a fruitcake to be made in the true English tradition, it must be meticulously and lovingly prepared by an experienced and appreciative fruitcake lover.

. . . After the fruitcake itself has been made, it is covered with a layer of marzipan, followed by a light and fluffy white icing of egg whites and confectioners sugar, which hardens to resemble snow. The cake is then decorated with Christmas scenes, using miniature wooden sleighs, plastic Santas, and the like.

Presumably, the diner can discriminate between the taste and texture of the cake and the plastic Santa.

But it is too easy to blame the English for what is now a worldwide problem. We are more concerned about the tight connection between fruitcake and alcohol. Timothy Taormino, of Baltimore, Maryland, admits to liking fruitcake, but is openminded enough to concede that "when it's bad, it's *BAD!*" What he may not realize is that all fruitcakes might taste bad if it weren't for the demon alcohol:

> I know of a recipe from Ireland that replaces the usual brandy or whiskey with Guinness Extra Stout. My girlfriend made it for a pot-luck dinner and it was quite a hit (especially when served with an Irish whiskey hard sauce).

Why do we get the feeling that dessert, or for that matter, the appetizer, was preceded by a few cocktails?

Similarly, Jack Adams, of Valencia, California, reports that the only fruitcake he ever liked was his grandmother's, and even this affection deserves a demurral:

> She bought a fruitcake from the store and would put a shot glass of whiskey in the center hole of the fruitcake. After a few weeks of this the cake became so saturated you didn't care what else was in it.
>
> Anyway, please don't publish my address. I've already got a shot glass and whiskey. That's all I need.

Jack seems to have the right idea. If you want to drink whiskey, cut out the middleman (i.e., fruitcake) and admit what you really want to consume. If you don't, you may end up like Nancy Schmidt, who not only admits to liking fruitcake but

> so loving its distinctive flavor that I purchase surplus loaves at the

DAVID FELDMAN

holidays to stock up so I can savor my favorite sweet yearlong.

Whew, now that I've publicly confessed to my fruitcake fetish, I'll either live a lauded life at the hands of other secret indulgers or, more likely, soon have uninvited guests in funny little white coats pounding at my front door.

Don't put yourself down, Nancy. Admitting your problem is the first step in solving it.

Honestly, now, despite the naysaying of the apologists, the sympathizers, and the fetishists, fruitcake truly is awful stuff. If people really do like fruitcake, why can't it compete on the open market? If anyone would ever order it, restaurants would offer fruitcake as a dessert. If fruitcake is so visually inviting and festive, why don't cafeterias ever offer it to lure customers? Wouldn't someone at Christmas dinner eat it (besides the baker of the cake, of course)?

We do not doubt the sincerity of the many readers who've had the courage to admit their dubious preference. We can only hope that greater minds than ours can someday finally figure out whether the preference is hereditary or environmental, mental or physical, spiritual or demonic. Until then, our mailbox is open to your theories, suggestions, and sordid confessions.

Submitted by Sheila Payne of Falmouth, Massachusetts.

A complimentary book goes to Nancy Schmidt of West New York, New Jersey, who perhaps will spend her holiday period reading instead of roaming the streets in search of surplus fruitcake; and to Melanie Morton, of Branford, Connecticut, in the fervent hope that reading this chapter together will help heal her family from the wrenching tragedy of fruitcake friction.

FRUSTABLE 2: *Why Does the Stroking of Index Fingers Against Each Other Mean "Tsk-Tsk"?*

We still don't have a definite answer to this Frustable, but two readers, Marsha Bruno of Norwich, Connecticut, and David

Schachow of West Hill, Ontario, came up with the identical theory. Although neither claims to have any evidence to prove the contention, it makes sense to us. Since David was the first to write, we'll quote him:

> The two index fingers are generally the two that are used in making the sign of the cross (and the same fingers we use to cross our fingers for good luck, or make the sign of the cross to ward off vampires and relatives).
>
> But why the stroking? Perhaps this is an evolved form of the whole cross (both fingers) being waved at or pushed toward the naughty-doer.

In other words, the "tsk-tsk" stroke is emblematic of pushing evil away.

Can anyone come up with anything better?

Submitted by Jim Hayden of Salem, Oregon. Thanks also to Mr. and Mrs. William H. McCollum of Oakdale, Minnesota.

FRUSTABLE 3: *We Often Hear the Cliché: "We Only Use 10 Percent of Our Brains." How Was It Determined That We Use 10 Percent and Not 5 Percent or 15 Percent?*

A few readers found written references to this cliché, but they have had no more luck than we did in tracking down its origins. Jeff White of Etobicoke, Ontario, and Albert J. Menaster of Los Angeles, California, both remembered that Richard Restak's 1984 book, *The Brain,* based on the PBS television series, mentioned the 10 percent theory. Menaster summarizes the contents:

> Restak says that the claim is probably based on studies showing large portions of the brain being damaged without any observable effects. . . . His conclusion is that since no one knows the number of neurons in the brain, it is simply impossible to determine how much of the brain is actually being used, and thus the 10 percent figure is without any basis and is unsupported by anything. I should add that I have read extensively on the subject of the brain, and I have never seen any scientific discussion of the 10 percent figure, which certainly supports Restak's position.

Restak goes on to note that the destruction of even a small portion of certain areas of the brain, such as the visual area, "can have a devastating effect."

One of the studies that Restak refers to obliquely is psychoneurologist Karl Lashley's, who removed portions of the cerebral cortex of rats without ruining their memory of how to run mazes. Reader Jeffrey McLean of Sterling Heights, Michigan, drew our attention to Carl Sagan's *The Dragons of Eden*, which discusses this issue. Sagan warns readers that just because we cannot see any behavioral change in a rat doesn't mean that there isn't a profound change when humans lose a portion of their brains:

> There is a popular contention that half or more of the brain is unused. From an evolutionary point of view this would be quite extraordinary: why should it have evolved if it had no function?

Sagan suggests that it is likely that the removal of a significant part of the brain does have a significant effect, even if we aren't currently capable of measuring or quantifying the change.

So, at least two popular science books testify to the existence of the cliché, but we are no closer to an answer to the genesis of the belief. Two friends swear that they remember reading about the 10 percent theory in a novel by Robert Heinlein, but we haven't been able to track it down yet.

Frustable 3 from *Why Do Dogs Have Wet Noses?* is still open for business.

*Submitted by David Fuller of East Hartford, Connecticut.
Thanks also to Ray Jackendoff of Waltham, Massachusetts, and
Jeff White of Etobicoke, Ontario.*

FRUSTABLE 4: *Where, Exactly, Did the Expression "Blue Plate Special" Come From?*

Reader Marty Flower provided us with a hot tip. She suggested we call the Homer Laughlin China Company, in Newell, West Virginia, the largest and one of the oldest suppliers of café and hotel china in the United States.

We called them and found to our consternation that although

they sell blue plates, they didn't start the practice, and didn't know the origin of "blue plate specials."

We were forlorn until we heard from Roger Bosley, of Arvada, Colorado, who sent along a reprint from a book (*A History of Man's Progress*) from and about Pioneer Village in Minden, Nebraska. This book claims that the now familiar blue willow pattern of china was inspired by a Chinese legend about a poor coolie, named Chang, who fell in love with Li Chi, the daughter of a mandarin, while playing under a willow tree. The mandarin forbade the relationship, and the willow tree drooped in sorrow over the broken romance.

The Chinese depicted the story on blue dishes, some of which were brought back to the West by Marco Polo. According to the book, written by Harold Ward, "restaurants serve their leading course on a blue willow plate and call it a 'Blue Plate' special—in tribute to this legend." Unfortunately we couldn't find any evidence connecting the "special" to the blue willow pattern.

We heard from a couple of people who encountered blue plate specials. Nazelle Trembly, of Ocean Grove, New Jersey, remembers that the plates had three-way partitions to keep sauces from running into one another. Trembly theorizes that these plates were exported from China, probably first used at sea, and then later shipped to port towns like New York, Boston, and San Francisco.

But Oree C. Weller, of Bellevue, Washington, is our only correspondent who ever washed a blue plate special. He believes that Americans imported this china from Japan!

> During the 1930's, Japan exported a lot of dishes, cups, and saucers in a hideous pattern. Cafés all over, especially the South, bought these dishes because they were cheap and the cafés could tolerate the high incidence of breakage by low-paid ($.10 per hour) dishwashers like me.
>
> The cafés served a fixed price, fixed menu lunch every day and soon customers began saying as they sat down for lunch: "What's on the blue plate for lunch today?" And hence the name stuck.

Perhaps we're no closer to knowing exactly where the expression comes from, but at least we have some tantalizing theories.

A complimentary book goes to Oree C. Weller of Bellevue, Washington.

FRUSTABLE 5: *Why Does the Traffic in Big Cities in the United States Seem Quieter Than in Big Cities in Other Parts of the World?*

All of our mail echoed the same sentiment: Traffic seems quieter in the U.S. because it *is* quieter. The hero, it seems, is the catalytic converter. Typical of the responses was the letter of Toledo, Ohio's David G. Conroy:

> Traffic in Europe seems louder because it is louder. The reason—emission standards. Since 1972, all cars made in America and those imported to America must have catalytic converters built into the mufflers. Not only do these little marvels clean up auto exhausts, but they also make cars quieter. If you disagree, simply take the shielding off the converter on your car and see how much noisier it becomes.

But other factors are involved, too, best summarized by reader Jerry Arvesen of Bloomington, Indiana:

> Our federal emissions laws are more stringent than most other countries'. Only recently are European countries requiring unleaded gasoline and the technology that reduces emissions, pollution, and consequently, noise.
>
> A better-running, state-of-the-art vehicle is much quieter than a carbon-belching monster. This is especially true when comparing the United States with countries from behind the Iron Curtain, whose technology in automobiles is the equivalent of the cars we were producing in the 1950s and 1960s.
>
> Also, the traffic laws of the United States are both enforced by police and obeyed by drivers much more than they are in European countries. This would naturally lead to less noise from horns honking, the sound of fender benders, drivers yelling at each other, and the like. I once read that in Italy, traffic lights are

a guideline. A green light there means to go without reservation; a red light means go, but look first.

Our cities and states are more modern and therefore laid out much more efficiently for vehicular traffic than they are in the older European cities that are centuries old and originally designed with narrow, twisting streets barely wide enough for horses and one-way traffic to pass through.

Submitted by Nityanandan Ashwath of Richmond Heights, Ohio.

A complimentary book goes to Jerry Arvesen of Bloomington, Indiana. Thanks also to David Schachow of West Hill, Ontario, and Ron Gulli of Tuscon, Arizona.

DAVID FELDMAN

FRUSTABLE 6: *Why Do Dogs Tilt Their Heads When You Talk to Them?*

We assumed that this would be the easiest of the ten Frustables to answer. We're still amazed that not one of the thirty or so dog experts we've contacted would venture an opinion on the issue.

Imponderables readers, however, have no such compunctions. Readers were split among three camps: those that thought the tilting had to do with the dog trying to hear better; those who thought the dog was trying to sharpen his vision; and the doggy anthropologists who are confident that the tilting is a sign of (pick one) aggression or friendliness.

Devotees of the last camp often compared dogs' tilting behavior to that of wolves, who are also known to tilt their head. Typical of the anthropology camp is the response of Marty Flowers, of Weirton, West Virginia:

> Dogs tilt their heads when you talk to them to let you know they are listening to you. They don't want to just stare at you, because that's a sign of aggression in the animal world, but if they look away it might seem that they are not paying attention to you. So they look at you but tilt their heads to show that it doesn't

mean aggression. Dogs don't growl and attack you with their heads tilted to one side.

We wouldn't know. We'd be too busy hightailing it away from the dog.

The eye-camp was best represented by Jim Vibber, of Tustin, California:

> Dogs aren't the only animals that tilt their heads when listening to humans talk, and I think this may relate to the answer.
>
> We humans often forget that most other animals do not perceive the world as we do. Binocular, 3-D vision probably should head the list of differences. Most animals (including dogs, birds, cattle, and fish) have one eye on each side of the head, and each eye sees half the world with little overlap in the fields of vision. We find it disconcerting to watch a chameleon looking at its surroundings, as each eye gawks around independently of the other like some clown doing cross-eye tricks. But we think it nothing unusual to watch a cockatoo turn its head sideways to get a close look at something. The same can be seen with goldfish and parakeets whenever you do something that gets their attention.
>
> Dogs and cats have eyes a little more forward on the head than, say, sheep or elephants, but not so far forward as people. They turn their heads sideways, but also frequently perform the more subtle movement of tilting the head at an angle while keeping the nose mostly pointed in the same direction. I've also seen this tilting movement in movies, such as when a wolf is looking at something, but I have no idea how much of this is in response to off-camera coaching by the animal trainer.
>
> This *may* be a way of looking at something tall, such as a human being or a tree. Has anyone checked whether dogs respond differently according to whether one is standing, sitting, or lying down?

Not to our knowledge. Before we write the next volume of *Imponderables,* we'll consult some more veterinary ophthalmologists and check out your theory, Jim.

One point that several dog experts emphasized to us is that dogs' hearing is so good that it is highly unlikely that they are tilting their heads in order to hear us. Still, we're most sympa-

thetic with the simpler but not unreasonable ear-theory propo-
nents, led by Susan Scott, of Baltimore, Maryland: "Wouldn't
you tilt your head if everyone around you were speaking gibber-
ish?"

We haven't given up yet. We're going to nail this Frustable
eventually.

*Submitted by Mark Seifred and Denise Meade-Seifred of
Memphis, Tennessee.*

*A complimentary book goes to Jim Vibber, who certainly has the
best rap, even if we're not sure we believe it.*

FRUSTABLE 7: *Why and Where Did the Notion Develop That "Fat People Are Jolly"?*

We didn't get much mail on this subject, but most of the letters
we did receive were choice.

Rick DeWitt, of Erie, Pennsylvania, sent us a reprint of an
essay written by Eric Berne (author of *Games People Play*) called
"Can People Be Judged by Their Appearance?," which first ap-
peared in his book *Mind in Action*. Berne argues that the three
main body types (endomorph, mesomorph, and ectomorph) each
yield specific personality characteristics.

According to Berne, the round, soft, thick build characteristic of the viscerotonic endomorph is usually possessed by someone who likes to "take in food, and affection, and approval as
well. Going to a banquet with people who like him is his idea of
a fine time." Berne's depiction of the endomorph is a catalog of
stereotypes about the jolly fat person ("The short, jolly, thickset,
red-faced politician with a cigar in his mouth, who always looks
as though he were about to have a stroke, is the best example of
this type.") with no evidence whatsoever to corroborate his con-
clusions.

Berne, a psychiatrist, notes that most people do not fall
clearly into one body type, but claims that if someone does, he
or she tends to display behavior characteristic of that body type

("If he is a viscerotonic, he will often want to go to a party where he can eat and drink and be in good company at a time when he might be better off attending to business . . ."). We're not sure Berne's discussion really answers our Frustable, but it surely demonstrates how pervasive the image of the fat, jolly person is.

We're more sympathetic with the homegrown theory of Kim Anderson, of Alma, Arkansas: "The excess fat under the skin of their faces hides wrinkles and stress lines so they appear to always be happy." This makes more sense to us than Berne.

But Melinda S. Mayfield, of Kansas City, Missouri, took us at our word about digging into the history of the fat/jolly notion:

> In ancient and medieval times, the physiologists believed that the four chief fluids or "cardinal humours" of the human body, blood, phlegm, choler (yellow bile), and melancholy (black bile), decided a person's physical and mental qualities and disposition by the dominance of one over the others. In the case of the humour blood, it created a temperament, or "complexion," called *sanguine*. A sanguine person was characterized by a ruddy countenance, a courageous, cheerful, amorous disposition, and an obese body.
>
> Even in William Shakespeare's day, people believed in the four temperaments, a fact evident in his plays. (Ever notice how the comic, happy people in them, such as Falstaff and Juliet's nurse, are fat? Well, now you know the reason.) In modern times, we no longer follow the theory of the four humours, but we do still follow Shakespeare's plays, and the idea of the florid-faced, jolly, roly-poly person has lived on.

Sounds pretty convincing to us.

> A *complimentary book goes to Melinda S. Mayfield of Kansas City, Missouri.*

FRUSTABLE 8: *Why Do Pigs Have Curly Tails?*

We spoke to many zoologists, veterinarians, and swine breeders about this topic and struck out, so we thought we would throw

DAVID FELDMAN

the question to readers. Lo and behold, your response was exactly the same as that of the "experts."

We heard from reader Nena Hackett, who used to raise pigs. She claims, as do the swine authorities we spoke to, that you can gauge the healthiness of the pig by the curl of its tail: "The tighter the tail, the less likely it will have parasites. If the tail is loose or just 'hanging around,' the pig will be sick every time." As Richard Landesman, associate professor of zoology at the University of Vermont, put it: "There seems to be only one good reason for the curl in a pig's tail, and that is to call the vet when it straightens out. More than likely, the trait for a curly tail is just part of the pig's genetic repertoire."

Maybe the uncurled tail is like the popout thermometer on a store-bought turkey. When you see it, it's nature's way of warning you to spring into action.

Submitted by Jill Clark of West Lafayette, Indiana. Thanks also to Colleen Crozier of Anchorage, Alaska, and George Hill of Brockville, Ontario.

A complimentary book goes to Nena Hackett of Harvey in the Hills, Florida.

FRUSTABLE 9: *Why Does the Heart Depicted in Illustrations Look Totally Different Than a Real Heart?*

Before we get too carried away with wild theories, our illustrious illustrator, Kassie Schwan, gently indicated that the "symbolic" heart doesn't look *that* unlike a real heart. The two upper lobes look a lot like real atria, and the "real" heart does taper at the bottom, although not as drastically as the heart we see on Valentine's Day cards and playing cards. From her point of view, the "symbolic" heart is much easier to draw than a "real" one.

So is our stereotyped heart merely the result of lazy efforts of mediocre artists? Anthropologist Desmond Morris, always quotable, if speculative, suggests in *Bodywatching* that the form of the symbolic heart might actually have been based on the shape of the female buttocks.

A startling theory? Not compared to the discovery of New York City broadcast designer Laura Tolkow, who was looking through a book of Egyptian hieroglyphics and stumbled across several upside-down stylized hearts depicted alongside a bird and a pyramid. Laura was shocked that our stylized conception of the heart dated back to ancient times, until she read the translation of the meaning of the upside-down hearts—they weren't hearts at all, but rather human testicles (right side up!).

So now we have our Valentine shape signifying the human heart, the female buttocks, and the testicles. Any other possible explanations?

We didn't think so until we heard from reader Howard Steyn, of Morristown, New Jersey, who claims that he was taught the answer to this Frustable in his seventh-grade science class! His teacher said that the reason for our stylized heart is the vessel structure surrounding the heart. Howard sent a diagram of the circulatory system of a frog, with a series of arteries, called the systemic arch, that looks *exactly* like the Valentine heart.

We immediately called up our favorite biologist, Professor John Hertner of Kearney State College in Nebraska, to talk to him about this breakthrough in Frustability. On very short notice, John conducted some comparative embryology and reported that indeed, most vertebrates, including humans, have a structural equivalent to the frog's systemic arch, although not necessarily the "perfect" heart.

Hertner made an important point that lends even more credence to Steyn's theory. In earlier days, the Catholic Church frowned on pathological or gross anatomical work on human bodies. Most European scientists conducting research on humans were thus not able to gain access to human cadavers. Many experiments were conducted on amphibians and rodents. There is a chance, in other words, that the systemic arch of a frog, or some other animal, was considered to be part of the heart, and perhaps even an assumption that the human heart looked like the frog's.

Submitted by Kathy Cripe of South Bend, Indiana.

A complimentary book goes to Howard Steyn of Morristown, New Jersey. Special thanks to John Hertner for help beyond the call of duty.

DO PENGUINS HAVE KNEES?

This question was inspired by our observation that everyone we have talked to thinks that they are a "net" loser of pens. So where do they all accumulate?

Evidently, a lot of them end up in Highland Park, Illinois:

> Most pens are probably lost through holes in pockets. I see lots of them lying on the ground, and pick up any that seem in good shape. I admit it! If you can prove that I have one of *your* pens, I would be happy to return it to you.
>
> Most of the rest of missing pens are probably borrowed temporarily, and not returned, accidentally.

This last thought was echoed by Bill Gerk, of Burlingame, California, who was the smoothest operator we heard from. With folks like Bill around, we know why banks chain their pens:

> I can't account for all of the missing pens. Just a few of them. Whenever I need to write a check in public or sign for a withdrawal, I ask the clerk or cashier, "May I borrow a pen?" After using the pen, I ask, "Did you give me this pen?" Usually the clerk or cashier will say, "Yes." I'll smile, say "Thank you," and start to put the pen into my pocket.
>
> But then I start to return it. About one out of ten times I hear, "That's O.K. You can keep it. We have a lot more." At first I did this for laughs, but if some choose to take me seriously, I settle for the pen, even if I don't get a laugh along with it.
>
> There are probably a few other similar, shameless, joking customers. Those who do this nefarious trick contribute to the disappearing pool of pens you're concerned about. By the way, if the pen you have been lent (given) doesn't write too well or you don't like the color, you may want to ask for another one before you sign anything.

We were surprised at how few readers sent theories about this Frustable. Our guess is that most readers, like us, are still losing pens and don't know why. Those of you in the state of Pennsylvania, however, are in serious jeopardy. We got a long, chatty letter from Philip M. Cohen, from West Chester, Penn-

DAVID FELDMAN

sylvania, that ended with these two chilling sentences: "Oh, one other thing. I have your pens." Aha!

The Frustables That Will Not Die

By their nature, Frustables aren't easy to solve. Even the crack *Imponderables* readers can't answer definitively some of our metaphysical quandaries. So we've promised you that we would keep you up to date on new contributions and discoveries in our search to take the Frust out of Frustables. This has become especially important since we've been putting out a new book of *Imponderables* each year, for it means that the readers of the paperback editions haven't had a chance to contribute to our forum. Here, then, are some of the best new ideas about old Frustables.

Frustables First Posed in *Why Do Clocks Run Clockwise?*

FRUSTABLE 1: *Why Do You So Often See One Shoe Lying on the Side of the Road?*

We have devoted more space to this topic than any other we have written about. We have received more mail about this subject than anything else we have ever written about. In *When Do Fish Sleep?*, we listed scores of theories to explain the phenomenon. Many of these theories assume that people deliberately throw shoes on the road. But until now, something was pointedly missing: an eyewitness account of a deliberate one-shoe toss.

We are no longer deprived. We heard from Joseph Metzelaar of Masonville, New York:

> I was reminded of an incident that happened to me while riding in a car driven by a woman wearing high heels. Her right foot repeatedly became wedged under the brake pedal, so out of sheer frustration she threw the right shoe out the window . . .

But don't get smug, readers. Cars aren't responsible for all one-shoe citings. And an alarming trend is evident. IT'S

SPREADING! We don't get too many letters from Sierra Leone, but we did get one from Peace Corps volunteer Jay D. Dillahunt, who is working in Freetown, Sierra Leone, West Africa:

> While walking down a bush path near my village, I spotted a single shoe lying in the path. There is no way it was tossed out of a car or bus window, because drivers of cars and buses have better sense than to drive down bush paths.

More proof that there is no escape from Imponderability.

FRUSTABLE 2: *Why Are Buttons on Men's Shirts and Jackets Arranged Differently from Those on Women's Shirts?*

Most readers seemed satisfied with the explanations we provided. But we heard from Tereen Flannigan of Livonia, Michigan, who said that she heard in school that during the Industrial Revolution, different taxes were imposed on the importation of men's and women's clothing in several European countries. Ingenious importers had the manufacturers change the button configuration to guarantee preferential tax treatment.

We haven't been able to confirm any of this. Does anyone else know more about this angle?

FRUSTABLE 9: *Why Don't You Ever See Really Tall Old People?*

Daphne Hare of Buffalo, New York, passed along a clipping from *Men's Health* magazine, with the results of an Ohio study that provides some pertinent data. In this study, men lost 1.2 years of life for every extra inch of height. That's right. A 6'0" man can expect to live six years less than a man of 5'7".

A previous study indicated even more dramatic height effects. Men who stood less than 5'8" lived to an average age of 82; those over six feet tall lived to the unripe age of 73.

DAVID FELDMAN

FRUSTABLE 10: *Why Do Only Older Men Seem to Have Hairy Ears?*

We're glad we put the word "Seem" in the question, for some younger men do have hairy ears. In fact, we heard from a middle-aged man who can so testify—Albert Jeliner of Mauwatosa, Wisconsin:

> I have had extremely hairy ears since I was in my late teens and early twenties. Back in those days it was long, blond fuzz and as the years have passed it has gotten more and more coarse. I am now 51. Since I was 22, I have had to have my ears trimmed each time I went to the barber.
>
> I don't know how you got into this "hairy" situation; but I inherited mine.

In *When Do Fish Sleep?*, we mentioned that all of the endocrinologists we spoke to begged off this question, claiming geneticists would have the answer (the geneticists, of course, sent us to the endocrinologists). But we heard from one endocrinologist, Dr. Clayton Reynolds of Lancaster, California, who claims the answer does lie within his field:

> The explanation for the hair growth in the ear canals of men is that these are the so-called androgenic zones. The hair follicles in the ear canals have receptors that are acted upon by the male hormones, or androgens.
>
> The hair of the human body can be divided into three categories: *nonsexual hair* (such as that on the arms and legs), which is not dependent upon the gender of the individual; *ambisexual hair*, which grows during adolescence in the same areas in the body in males and females (such as the axillary and pubic hair); and thirdly, *male sexual hair*, which grows on the face and chest and between the umbilicus and the pubic area.
>
> Not so well known is the fact that the male sexual hair occurs also in the other androgenic zones, the ear canals.

But why would this hair tend to sprout as we get older? We heard from University of Texas medical student John Chaconas, of Corpus Christi:

There are two types of hair on the body, *vellus* and *terminal*. Vellus is present on the parts of the body usually considered hairless, such as the forehead, eyelids, and eardrums. Terminal hair is present on our heads and arms.

Sometimes, as we age, vellus hair follicles are differentiated (transformed) into terminal follicles, and give rise to terminal hair. This transformation usually does not occur at an early age—therefore only older man have hairy ears.

Frustables First Posed in *When Do Fish Sleep?*

FRUSTABLE 3: *How, When, and Why Did the Banana Become the Universal Slipping Agent in Vaudeville and Movies?*

In *Why Do Dogs Have Wet Noses?*, we announced that we had made no progress at all in answering this Frustable. We're now a little, and we stress the word "little," further along in our quest. Most of our mail assumes that the banana's role as a slipping agent came from burlesque itself. Peter Womut of Portland, Oregon, makes an intelligent stab:

> I believe the connection is with the three comedians in a burlesque show and their title of top, second, and third bananas. Perhaps they became known by these names because of slips on banana peels, or perhaps banana peels became the symbol because of their names. Perhaps the solution is in the lyrics to the song, "If You Want to Be a Top Banana," from the musical *Top Banana*.

After an appearance on "CBS This Morning," when we mentioned our frustration about solving this mystery, we received a call from an excited law student who promised to but never did send us information about some cases he was studying in torts class. Lo and behold, the Supreme Court, and Oliver Wendell Holmes himself, rendered several decisions on personal injury cases involving folks slipping on banana peels on the street.

DAVID FELDMAN

Why were there banana peels on the street? The law student cited the same reason that Les Aldridge, of Milltown, New Jersey, wrote to us about:

> My wife reports that her grandfather used to say that when he was young, people would often use banana peels (the soft inner part) to shine their shoes. If they got some of it on the bottom of the shoe and fell, they would say that they had "slipped on a banana peel."
>
> The banana peel makes an easily recognizable prop for the stage to give comedians the occasion for humorous pratfalls. They needed some slipping agent—an egg might do it but it would be too messy. Perhaps the banana peel suggested itself as the agent due to the shoe shining association mentioned above.

Shannon Arledge, of Livingston, Louisiana, adds an even stranger use for the lowly banana peel: "In olden times, banana peels were often used on boat docks as a way of sliding the boats down to the water. However funny, slipping on those peels was a common occurrence." We'll try to track down the Supreme Court cases on banana peels for our next opus.

FRUSTABLE 6: *Why Do So Many People Save* National Geographics *and Then Never Look at Them Again?*

We thought we had covered all the possible answers to this Frustable in *Why Do Dogs Have Wet Noses?*, but a letter from Laurie Poindexter, of Fowler, Indiana, made us realize we had been negligent:

> My mother used *National Geographics* as booster seats for all of us during childhood, so we could easily reach the table at mealtimes. Once we had all grown sufficiently, we sold them at yard sales.
>
> Now that she has grandchildren, she is subscribing to *Smithsonian*, which has equal thickness but a wider base for more secure stackability and bottom comfort.

Yes, but do they come with lap restraints and air bags?
But we shouldn't joke about this matter. Mark Arend of Bea-

ver Dam, Wisconsin, was kind enough to send us an article by George H. Kaub, published in the *Journal of Irreproducible Results*. This ground-breaking article, entitled, *"National Geographic, The Doomsday Machine,"* succinctly states the problem:

> Since no copies have been discarded or destroyed since the beginning of publication it can be readily seen that the accumulated aggregate weight is a figure that not only boggles the mind but is imminently approaching the disaster point. That point will be the time when the geologic substructure of the country can no longer support the incredible load and subsidence will occur.

Kaub concludes that the entire country will fall below the sea and "total inundation will occur" unless drastic measures are taken. *National Geographic,* he is quite sure, is responsible for many of the so-called natural calamities of our time—including activity on the San Andreas fault.

Kaub's article is a clarion call, warning us that unless the publication of *National Geographic* is terminated, the North American continent is doomed.

FRUSTABLE 8: *Why Do Kids Tend to Like Meat Well Done (and Then Prefer It Rarer and Rarer As They Get Older)?*

Several readers have written in to add a third possible answer to the two we tentatively suggested. Most persuasive was Babette Hills of Aurora, Colorado: "I asked my two boys why they don't like their meat rare and their answers were just what I had observed: Rare meat is harder to chew." Rare meat, still containing more natural juices and fats than a well-done piece, is chewier. Hills argues that "baby" teeth are not equipped to do the necessary damage to a piece of rare meat with ease or comfort. This also explains why children often like to have their meat cut into small pieces.

Hills assumes that as soon as chewing the meat is no longer a problem, most kids will start preferring rarer meat, which is more flavorful. Her theory would be confirmed, she says, if people with missing teeth or dentures like their meat more well done than those not dentally challenged.

FRUSTABLE 10: *Why Are So Many Restaurants, Especially Diners and Coffee Shops, Obsessed with Mating Ketchup Bottles at the End of the Day?*

We thought we answered this Frustable definitively. But then we received a startling new reason for this seemingly silly obsession: Ketchup mating saves lives. Waitress Mary A. Taft of Deming, Washington, explains:

> As a waitress, I had often asked myself, "What is the object of this tedious chore?" One day, I finally got my answer when a bottle of ketchup exploded when the lid was taken off.
>
> My boss informed me that to prevent explosions of this kind, the bottles low in ketchup must be poured on top of the bottles more full of ketchup. Otherwise, some ketchup can end up perpetually on the bottom of the bottle instead of on someone's burger.
>
> If the ketchup stays on the bottom long enough, it can ferment, causing a buildup of gas that can create the explosion.

Of course, this explosive effect can better be cited as a reason to eliminate the practice of mating altogether. But we won't hold our breath until restaurateurs change their practice.

Tuesday
Dear Dave ~
I believe I
have the
nswer to
24815
US MAIL
LETTERS

We receive thousands of letters every year. Most of them pose Imponderables, try to answer Frustables, or ply us with undeserved praise. But this section is reserved for those letters that take us to task or have something significant to add to our statements in previous books.

Many of you have written with valid objections to parts of our answers in previous books. As much as it pains us to make a mistake, we appreciate your input. Unfortunately, it can take months or even years to validate these objections. We have made several changes in later printings of our books because of your letters. The letters we include here are not the only accurate comments, just some of the most entertaining.

In Why Do Dogs Have Wet Noses?, *we discussed why there is no Boeing 717. We heard from a pioneer of the Boeing Company, who told us that the version we related was a little sanitized:*

> I was assigned to the group that prepared the detail specifications for the various airplane models. We put together a Boeing Model 717 detail specification and used it in preliminary configuration negotiations with United Airlines.
>
> William Patterson, who started his aviation career as William E. Boeing's administrative assistant in the 1930s, was president of United Airlines at the time. He was very pro-Douglas Aircraft, and had stated that he would never buy a Boeing Model 707 and that "Model 717 was too close to 707."
>
> We obliged him by changing the name to "Boeing Model 720" and he (and many other airlines around the world) bought it. That's the only gap in the "7–7 Series." All we had to do to the specification was change the title page. When we sold the 720 to UAL, no engineering drawings had been released, so other than my small wages, the change cost Boeing nothing. End of story.
>
> OREE C. WELLER
> *Bellevue, Washington*

In Imponderables, *we discussed why cashews aren't sold in their shells. We blew the answer. We indicated that cashews are actually seeds and that they don't have hard shells. Well, the cashews we eat are seeds, but M. J. Navarro of Latrobe, Penn-*

sylvania, humbled us by sending some unshelled cashews that proved our discussion erroneous. He added: "The shell is too hard for people to have any chance at the seed itself. The shell must be thoroughly roasted to get to the seed and the seed must be roasted to make it edible."

True enough. And there are other reasons why cashews are not sold unshelled in stores. We heard from a reader who spent more than a little time with cashew shells during his childhood:

> The mature shell is thick and leathery and contains a corrosive brownish black oil that causes blisters on skin. . . . The juice of the apple causes indelible stains on clothes . . .
>
> It would be uneconomical to import cashews in their shells, and that is why they are not sold this way . . .
>
> As a child, I grew up with cashew trees all around, played with thousands of unshelled cashews (a game similar to marbles), raided the cashew grove near my high school, and watched moonshine cashew liquor being made as well as sold and consumed.

> DINESH NETTAR
> Edison, New Jersey

And a special apology to Daniel Pittman of LaSalle, Illinois, who defended our discussion of cashews to his merciless friends.

In Why Do Dogs Have Wet Noses?, we discussed why a new order of checks starts with the number 101 rather than 1. Walter Nadel of Pembroke, Massachusetts, provides a technical reason why starting with 101 saves checkmakers some money:

> When checks are printed, the first sheet through is at the bottom of the hopper, and the last sheet through is at the top. This requires backward-functioning numbering machines, which are set to start at the ending number and finish with the starting number. The order of checks can then be completed and shipped in the proper numerical sequence ready for use by the customer.
>
> As the order is printed, and the numbers recede, they would read 102, 101, 100, 099, 098, etc., unless the press is stopped and the superfluous zero is manually depressed after 100. Again, at the

DAVID FELDMAN

point of 12, 11, 10, 09, the press would have to be stopped again to eliminate the second superfluous zero.

By encouraging the use of starting check orders with 101, the printer has saved the need to stop the press twice, a saving in time and efficiency, not to mention money.

WALTER NADEL
Pembroke, Massachusetts

Many readers were not satisfied with the explanation provided in Why Do Clocks Run Clockwise? *about why IIII, and not IV, is used to signify the number four on clocks. We heard from Alden Foltz of Port Huron, Michigan, who collects ancient coins. Foltz assures us that all of his Roman coins used "IIII" to signify the number four. If coins don't reflect the proper way to express four, what would? asks Foltz.*

Further corroboration comes from Bruce Umbarger, who, like many readers, wondered how our explanation resolved the dilemma of how 9 would be expressed in Roman numerals in ancient times. Bruce consulted a book called Number Words and Number Symbols, *by Kurt Menninger, which clearly shows "IIII" used to express 4 from as early as circa 130 B.C. He also enclosed two multiplication tables from thirteenth-century monastic manuscripts: They also expressed 4 as "IIII." "How can the habit of 'IIII' on clocks be traced to illiterate peasants when a copy of a sixteenth-century astrolabe [a device used to tell the time by observing the stars or sun] clearly shows a four rendered as 'IIII'?"*

So why might the IIII have been changed, eventually, to IV? Reader Tom Woosnan of Hillsborough, California, sent us an excerpt of Isaac Asimov's Asimov on Numbers *that provides one possible explanation:*

> [IV] . . . are the first letters of IVPITER, the chief of the Roman gods, and the Romans may have had a delicacy about writing even the beginning of the name. Even today, on clock faces bearing Roman numerals, "four" is represented as IIII and never IV. This is not because the clock face does not accept the subtractive principle, for "nine" is represented as IX and never as VIIII.

An alternate explanation is that the expression of Roman numerals was not standardized throughout the Roman Empire, let alone throughout Europe.

Speaking of clocks, we heard from someone who was presented with a copy of Why Do Clocks Run Clockwise? *for his eighty-third birthday:*

> The title struck a tender chord with me, for I was asked to reset a friend's wall clock last fall, when folks were going from daylight savings time to the daylight wasting time.
>
> I had pulled the clock's cord from the wall outlet to make the change. When I returned the connection in the wall's electric outlet, the clock was running backwards!
>
> I pulled the cord from the outlet, turned the male fitting of the cord over, and plugged it in again. It ran forward, properly.
>
> Why? God only knows. And he won't tell.
>
> VERNE O. PHELPS
Edina, Minnesota

For some reason, we tend to make exactly one really *stupid mistake in each book. In* Imponderables, *it was the cashew question, but in* When Do Fish Sleep?, *it was this one:*

> For shame, young man . . . Crickets do not produce chirps by rubbing their legs together. They have on each front wing a sharp edge, the scraper, and a file-like ridge, the file. They chirp by elevating the front wings and moving them so that the scraper of one wing rubs on the file of the other wing, giving a pulse, the chirp, generally on the closing stroke.
>
> On a big male cricket, the scraper and the file can often be seen by the naked eye. You can take the wings of a cricket in your fingers and make the chirp sound yourself.
>
> CLIFFORD DENNIS
PH.D. ENTOMOLOGY
Prairie du Chien, Wisconsin

In Why Do Clocks Run Clockwise?, *we discussed why dogs circle around before lying down. Although we stand by our answer, several readers wrote to us with an additional possible*

DAVID FELDMAN

reason: to determine which way the wind is blowing. In this way, they can sleep with noses into the wind, in order to be warned of a predator approaching. Being on the "right" side of the wind may also help them hear predators.

In Imponderables, *we discussed who is criminally responsible when an elevator is illegally overloaded. One reader indicates that the answer might be . . . a restaurant:*

> A few years back, I was in a building in Washington, D.C., with passenger elevators that might be accurately described as cozy. The building had an Italian restaurant on the ground floor, and as I arrived, four secretaries from an upstairs office were just getting on the elevator to go back to work, having had lunch together downstairs.
>
> Every time they tried, though, the same thing happened: the doors would close, but as soon as they pressed the button for their floor, the doors would open again. They became rather angry at me, thinking I was doing something to keep the elevator from working properly.
>
> I finally suggested that it appeared to me that they had overloaded the car and that it would probably work if one of them got out. One did leave the elevator, and the other three women were able to return to their floor immediately.
>
> What amused me, though the women were rather embarrassed about it, was that the four were able to ride down together for their lunch, but their group weight gain during the meal was enough to trip the overload protection.

DALE NEIBURG
Laurel, Maryland

On to a less weighty subject, we heard from Terri Davis, of Newark, Delaware . . .

> Reading your book *Why Do Clocks Run Clockwise?*, I came across the question, "Why do so many mass mailers use return envelopes with windows?" The answer was interesting and reminded me of something I saw at work last year that answered a question I never thought to ask: What happens to the paper that gets punched out of envelopes to provide the window?
>
> Scott Paper, Inc. uses a great deal of recycled fiber during the manufacturing process of its many products—so much so that it

buys scraps from many other companies . . . Can you imagine the sight of 125 cubic feet of compressed envelope windows heading down to a watery grave? There were literally tons more of these windows waiting on the incoming barge to be turned into toilet paper, napkins, tissues, etc.

We're happy that at least one company is making good use of scrap paper. After all, in Why Do Clocks Run Clockwise?, *we already chastised doughnut stores for having clerks picking up doughnuts with tissues and then stuffing the same tissues, germs and all, into the bag the customer takes home. Jack Schwager of Goldens Bridge, New York, mounts a passionate defense of the practice:*

> Germs are not the only issue. If they were, then indeed it would make no sense to put the tissue in the bag. Let me suggest a simple experiment, which would make it clear to anyone why it would be desirable to leave the doughnut wrapped in a tissue.
>
> Wait for a hot, humid summer day. Go into Dunkin' Donuts and order a dozen doughnuts. Make sure they all have icing and that the attendant uses a bag instead of a box. Now go into an unairconditioned car (preferably black in color) and drive around, taking care of several errands before returning home. Upon returning home, separate the doughnuts, giving the one with chocolate icing to the anxious child who loves chocolate but hates strawberry, and the strawberry one to his brother, who hates chocolate.
>
> What? You can't neatly separate the doughnuts, which have by now turned into a congealed blob? Hmmmm.

Hmmm, indeed. It has been our experience that the tissues do absolutely no good in keeping the frosting separated, since they aren't large enough to surround the doughnuts and the tissues are far from surgically inserted to minimize friction between icings. We won't even go into the dreaded powdered sugar crisis, in which every doughnut in the bag becomes sprinkled with the white stuff whether you like it or not. If the stores really wanted to make sure the doughnuts did not mate, they could certainly make a form-fitting tissue for them, a little baggie that covered the pastries, or provide a box with dividers.

Readers are still fuming about the purpose of the balls on top of flagpoles. Everyone seems totally sure of the purpose, but we've gotten about ten different explanations. We've received a few letters like this one, from Heather Muir of Phoenix, Arizona:

> In the military, the ball on the top of the flagpole is referred to as a "truck." The "truck" contains a gold bullet, a silver bullet, and a match. Buried at the base of the flagpole is a 50mm shell. The bullets and the shells are essentially memorials, in remembrance of the various wars the United States has been engaged in. The match, however, is stored in the "truck" for one purpose—to burn the flag in case the post is overrun by enemy troops, thus preventing the desecration of our national symbol.

Two problems with this explanation. First, the representatives of the armed services we spoke to say that it isn't true. And second, if your post is being overrun, would you really have time to take off the truck and retrieve a match? Wouldn't matches be more readily available elsewhere?

We heard from Elsie McManigal of Hercules, California, who said that before World War I, when most flagpoles were made of wood, balls were inserted to prevent water damage to the open grain on the top of the flagpole.

Betsy Kimak of Boulder, Colorado, took us to task for our discussion of why two horses in an open field always seem to stand head to tail. She notes that horses, and other ungulates, use posture to indicate who is the dominant animal:

> With only two horses alone in a field, one will always be dominant and aggressive, depending on size, strength, and sex.
>
> Two different postures are applicable here. The first is initiated by the submissive horse, who faces away from the dominant one to announce that it is not a threat and has no intentions of fighting or challenging the dominant horse. The second posture is initiated by the dominant horse and is called the "dominance pursuit march." The dominant horse crowds in behind the second horse and initiates a march head-to-tail, in which the recipient horse must accept submissiveness or face an aggressive confrontation.

DO PENGUINS HAVE KNEES?   241

We heard from Bill Beauman, research manager at Ever-pure, Inc., who wanted to add to our discussion in When Do Fish Sleep? *about why some ice cubes come out cloudy and others come out clear.*

At home, using still (not moving) water in an ice cube tray in the freezer, air entrapment may be a minor cause of cloudiness, but the most common explanation is that the salts and other minerals in nearly all water supplies are gradually forced to the center of the cube by the freezing of the "pure" water around the sides. As the concentration rises and exceeds the minerals' natural limits of solubility, they precipitate as particles and make an ugly, cloudy center. When the ice melts, the minerals make sediment in the glass.

Commercial cube-type icemakers recirculate water from a sump to cascade over freezing plates. Both dissolved and particulate impurities are readily rinsed away, and only the relatively pure water freezes. Such equipment routinely makes ice which, when melted, contains only about 5 percent of the minerals originally present in the water [so this is why ice cubes often don't taste like the water they are made out of!]. . . . Such machines always make clear ice, unless there isn't enough refrigerant to make the plates cold enough.

What is an Imponderables *book without a ketchup controversy? We heard from the president of The Throop Group, William M. Throop, who questioned Heinz's explanation for the necessity of the neck band on their bottles (they said it was a way to keep the foil cap snug against its cork and sealing wax prior to the introduction of the screw-on cap in 1888).*

Sealing wax sticks quite well to bare glass. The real reason for the neck band is that air left in the bottle allowed a black ring of oxidized material to form. The neck band was designed to hide this black ring. Although it did not detract from the product quality, the ring was unsightly.

In 1952, the company for which we worked (Chain Belt Co., now Rexnord Inc.) developed a product called a Deaerator. This removed the air pocket from the bottle when filling and eliminated this black layer at the top of the bottle. The neck band was retained as a signature of Heinz ketchup. However, all other

DAVID FELDMAN

brands of ketchup used similar neck bands prior to deaerating for the same purpose.

Speaking of packaging, we are amazed that we actually heard from someone who defended milk cartons (we wrote about why they were difficult to open and impossible to close in Imponderables*):*

> Your assertion that paper milk cartons are "difficult to open ... and close" goes far beyond reality. After a few initial experiments years ago, I found opening and closing such paper containers rather simple.
>
> One need only place the side to be opened facing one, insert a thumb on either side of the rooflike structure under the eaves, twist one's hands forward and upward, steadying the container with the fingers until the roof portion pushed by the thumbs becomes detached from the roof peak and pressed forward by the thumbs until nearly flush against the carton top. Then hold the carton with one hand while using the other thumb and one or two fingers to pinch back the edges of the eaves the thumbs have just pushed back, and bring them forward, which forms an opening from which to pour the milk.
>
> To close, merely reverse the process. The whole operation should take three to five seconds by the third attempt. No problem!

> JERROLD E. MARKHAM
> *Hansville, Washington*

We rest our case.

Jerrold, do you by any chance know Jan Harrington of New York, New York? If not, we think you have the potential for a wonderful friendship based on mutual proclivities. In Why Do Clocks Run Clockwise?, *we whined that Coca-Cola from two- or three-liter bottles just doesn't taste as good as from smaller containers. Jan Harrington agrees with us that large bottles have too large an air capacity, but Jan, unlike us, does something about it:*

> I never buy the three-liter (they don't fit in the refrigerator door and never cost less per liter than the two-liter bottles), but I

have a solution for the two-liter bottles. They are flexible. Once opened, I squeeze the plastic so that the liquid fills the entire volume of the container and keep squeezing more and more air out before closing the cap each time I pour a later drink. I find this preserves the flavor much better than leaving the bottle as a bottle shape.

Try it yourself (if you have small hands you can press the bottle against your chest using one arm until the Coca-Cola has just filled the bottle, then screw the cap on tightly). The bottle will retain the squozen shape. Eventually, you have a plastic bottle that looks as if run over by a car tire.

Heaven knows what Jan does with toothpaste tubes. We prefer just whining about the Coke problem.

Speaking of liquids, we have another theory about why the cold water that comes from the bathroom faucet seems colder than the cold water from the kitchen faucet:

Most kitchen faucets are equipped with water-saving aerators, which decrease the percentage of water in the stream; the stream from the kitchen faucet is actually part water *and* part air at room temperature and therefore can not be as cold as the stream from the bathroom faucet. Even kitchen water that is dispensed into a glass has been slightly warmed by this phenomenon.

Furthermore, the body parts that encounter water in the bathroom (face and body) are far more unaccustomed and sensitive to cold than the parts that encounter water in the kitchen (usually only hands).

Tony Acquaviva
Towson, Maryland

O.K., but the original impetus for the question was that several readers went to the bathroom to get drinking water because it was colder. And we have experienced the same phenomenon in kitchen faucets without aerators.

Speaking of the kitchen, Rice Krispies are often eaten there:

In *When Do Fish Sleep?*, you told us how Rice Krispies go Snap! Crackle! and Pop! Now could you explain why in French

DAVID FELDMAN

they go Cric! Crac! Croc!? And in Japanese, they go Pitchi! Patchi! Putchi!?

M. PICARD
Montreal, Quebec

No, M. Picard, we can't.

Speaking of complex carbohydrates, Ivan Pfalser of Caney, Kansas, took us to task for not including another reason why silos are round in When Do Fish Sleep?:

It's the same reason that water towers, grain elevators, gas tanks, and barrels are all round. A circle is the most efficient configuration to carry the stress produced by an internal load. . . . If a square or straight-sided silo is used, the internal pressure tries to bend the side into a hoop shape, inducing a momentary bending . . . which can cause overstressing in the corners and can cause them to tear apart.

Speaking of tearing apart, the issue of "Where are the missing socks?" is evidently ripping apart the fiber of the Western world. We received letters from smug individuals who pin their socks together when they throw them in their laundry hamper and claimed they never lose a sock. In our opinion, they are merely shrugging off the challenging and character-building exercise of trying to preserve the sanctity of sock coupling. We were a little ironic in our treatment of this issue in Why Do Dogs Have Wet Noses?. *So we'll let an expert, Sam Warden of Portland, Oregon, who describes himself as a "multifarious mechanic and not-bad-housekeeper-for-a-guy," provide pretty much the same answer we do when confronted with this question by investigative reporters and talk-show callers:*

Yes, Virginia. Socks do abscond in the wash. I have watched plumbers pull socks out of blocked laundry drains; I have done so myself. I have retrieved them in midflight from my laundry sink, and found them (not surprisingly) inside jammed washing machine pumps. Every plumber and repairman wherever washing machines are used must know about this.

The great majority of fugitives are children's sizes and short-length sheer stockings. (Handkerchiefs jump ship, too; they just

DO PENGUINS HAVE KNEES?

aren't kept in pairs.) The cheaper models of washing machines and front loaders seem to be the main offenders.

And may we add the other major culprit. We have found many a sock clinging fervently to the drums of dryers.

And in reference to clinging of the static variety, several readers wrote to us about an additional reason why television sets are measured diagonally (we discussed this Imponderable in Why Do Dogs Have Wet Noses?*):*

> The early TV picture tubes were all circular, so that the size of the tube was the diameter of the circle. A rectangle was masked on the front; thus, the diameter of the tube became the diagonal of the picture. In order to make the picture as large as possible, the corners of the picture were curved.
>
> When the first rectangular tubes were produced, they were measured on the diagonal, since this was equivalent to the diameter of the round tubes that they were meant to replace.
>
> ROY V. HUGHSON
> *Stamford, Connecticut*

Speaking of technological devices throwing off light, Douglas S. Pisik of Marietta, Georgia, read our discussion in Why Do Clocks Run Clockwise? *and says that he learned in driver education (and you know how reliable driver education teachers are) why the red light is on top of a traffic signal—if you are coming over a hill, you will see the red light first. This won't help you pass a driver's test, but it makes a little sense.*

We don't usually discuss letters about our word book, Who Put the Butter in Butterfly?, *in these pages, but we couldn't resist sharing the information we were seeking about the origin of our favorite English euphemism for sex, "discussing Uganda":*

> "Discussing Uganda" was popularized in England following an incident at Heathrow Airport when the (female) Ugandan ambassador was caught using the restroom for uses other than those for which it was intended [how's *that* for a euphemism].
>
> *Private Eye,* a British magazine dedicated to humorously exposing such stories, ran this piece and then continued to discuss

DAVID FELDMAN

similar incidents using the euphemism, "discussing Ugandan relations," helping the magazine avoid many potential lawsuits from those who were exposed.

MARTIN BISHOP
Marina del Rey, California

And what would a letter column be without the "mother of all Imponderables," which we first discussed in Why Do Clocks Run Clockwise? *and then have discussed in every subsequent* Imponderables *book: Why do some ranchers hang old boots on fenceposts?*

Since we first wrote about the subject, it has become en vogue among the literary set. Lance Tock of Brooklyn Park, Minnesota, sent us an excerpt from a book called Farm, *written by Grant Heilman, that endorses one of the theories we've mentioned—that boots were put on top of fences to prevent rainfall from rotting post ends. We received a copy of a newsletter called* Nebraska Veterinary Views *that includes a regular column by Dr. Larry Williams entitled "Old Boots & Fence Posts." He was shocked to find that only half of a group of forty Nebraskans he had spoken to had seen old boots on fenceposts, "let alone wonder[ing] about the[ir] meaning. They claim to be Cornhuskers, but can they be 'real' Nebraskans until they have seen and held in reverence old boots and fence posts?"*

Paul Kotter of Lebanon, New Hampshire, informed us that in Tony Hillerman's novel Talking God, *page 28 contains the sentence, "An old boot was jammed atop the post, signaling that someone would be at home." We don't buy this theory, because the boots observed in Kansas and Nebraska, anyway, stay on the posts all the time.*

Duane Woerman, who lives in Kearney, Nebraska, near the epicenter of boot-fencepost activity, has strong opinions:

> In Nebraska, the shoe is placed upside-down so that the cowboy's *souls* [get it?] will go to heaven. The toes of the boots are always pointed towards the house. In a snowstorm, the cowboys can always find their way home by following the direction in which the toes are pointed.

DO PENGUINS HAVE KNEES?

247

If it's that bad a storm, the cowboys may end up at someone else's home, but we get the point. At least a poet with the same answer admits that there are many possible explanations. Roger Hill of Vandalia, Ohio, sent us a poem called "Roadside Riddle" by Faye Tanner Cool, who writes about the boots along the prairie highways near her home of Fleming, Colorado. The last five lines of the poem:

> The local folk
> offer up two dozen explanations,
> but the one to ponder on:
> as reward for miles of walking
> those soles face heaven.

Veterinarian Lucy Hirsch of Smithville, Missouri, suggests that shoes and boots are put on metal fenceposts for a most practical reason: "Metal fenceposts known as 'T' posts are sharp. Many horses have impaled themselves on them. Boots and shoes on top can protect the horses from the sharp points." *But they do things differently in South Carolina. Amanda Stanley writes to inform us that we are wrong to be worrying about the fenceposts. We should be worrying about the boots:*

> In South Carolina, ranchers hang them to dry the leather in the sun after a big rain. After the leather has dried, they grease them down with mink oil. When old boots get hard, they leak water; after they are dried, the mink oil will stop the leaks. The smell of the boots will keep crows out of the fields.
>
> . . . I hope this answer will get this Imponderable out of the Frustables section.

It's way too late for that.

Finally, we have always feared that Imponderables *might be soporific. We never dared dream they could be used as aphrodisiacs. We received a letter from a woman from Burbank, California, whose name we'll keep anonymous, which begins in the following way:* "Every night, my lover reads to me from Why Do Clocks Run Clockwise?." *Now there is a basis for a wonderful relationship.*

DAVID FELDMAN

Acknowledgments

The best part of my job, even better than cashing my paychecks, is hearing from you. Not only do readers supply most of the Imponderables, but your encouragement and support have kept me going at times when I've wanted to pack it in. More than anyone, you deserve my gratitude.

I've tried to express my thanks by responding personally to readers who enclose a self-addressed stamped envelope, and I will continue to do so. But I can't promise I'll be prompt in my response. *Imponderables* receives thousands of letters a year now, so if I'm on deadline or on the road promoting the books, I fall behind in my correspondence. I apologize for the delay but hope you agree that the alternative (form letters) just wouldn't be as satisfying for you or me. Rest assured that I treasure and read every letter I receive.

This is my fifth book at HarperCollins, and luckily for me, my fifth with Rick Kot as editor. Since the last book, Rick has been promoted once again—and he has no relatives in upper management to explain this inexorable rise, either. Rick's assistant, Sheila Gillooly, has been a godsend, dispatching problems with unfailing intelligence and good humor. My publicist, Craig Herman, has been responsible for thrusting *Imponderables* onto the airwaves and his assistant, Andrew Malkin, has been an able and enthusiastic co-conspirator. And the production editing/copy editing team of Kim Lewis, Maureen Clark, and Janet Byrne helped make this book at least semicomprehensible.

Thanks to the many movers and shakers at HarperCollins who have allowed me to concentrate on the writing while they worried about the flogging: publisher Bill Shinker; Roz Barrow; sales honchos Brenda Marsh, Pat Jonas, Zeb Burgess, and all the HC sales reps; marketing mavens Steve Magnuson and Robert Jones; special marketing titans Connie Levinson and Mark Landau (and all my other pals in special markets); and primo publicist Karen Mender. And thanks to all the folks at HarperCollins with less lofty titles, who have been so kind and supportive when they didn't have to be.

Jim Trupin, wise beyond his years, as he will be only too glad to inform you, is a terrific agent and friend, as is his wife and partner, Elizabeth.

If Kassie Schwan wrote as well as I drew, she'd be unreadable. So you will understand why I so appreciate her cartoons; if you knew Kassie, you'd certainly understand why I prize her so much as a friend.

Lest I be accused of Dorian Grayish tendencies, a team of thugs lugged me in front of a camera to finally change the photograph on the back cover. I wouldn't have sat still in front of a camera for a minute without the gentle coaxing and sensitivity of photographer supreme, Joann Carney. Thanks to Fran Hackett and Peter Fenimore, of the New York Aquarium in Coney Island, for introducing me to Klousseau, the very special penguin featured on the cover. James Gorman, author of *The Total Penguin*, was kind enough to lead me to the folks who would allow me the privilege of spending an hour with Klousseau and his feathered colleagues.

Several of my personal friends just happen to work in publishing, and have been invaluable sources of support and counsel and, more often than I would like to admit, passive and helpless recipients of my endless whining. Thank you Mark Kohut, Susie Russenberger, Barbara Rittenhouse, and James Gleick.

And then there are my friends and family, who see me disappear for months at a time during deadline sieges or publicity tours. Thanks for putting up with me to: Tony Alessandrini; Michael Barson; Sherry Barson; Rajat Basu; Ruth Basu; Jeff Bayone; Jean Behrend; Brenda Berkman; Cathy Berkman; Sharyn Bishop; Carri Blees; Christopher Blees; Jon Blees; everyone at Bowling Green State University's Popular Culture Department; Jerry Braithwaite; Annette Brown; Arvin Brown; Herman Brown; Joann Carney; Lizzie Carney; Susie Carney; Janice Carr; Lapt Chan; Mary Clifford; Don Cline; Alvin Cooperman;

DAVID FELDMAN

Marilyn Cooperman; Judith Dahlman; Paul Dahlman; Shelly de Satnick; Charlie Doherty; Laurel Doherty; Joyce Ebert; Pam Elam; Andrew Elliot; Steve Feinberg; Fred Feldman; Gilda Feldman; Michael Feldman; Phil Feldman; Ron Felton; Phyllis Fineman; Kris Fister; Mary Flannery; Linda Frank; Elizabeth Frenchman; Susan Friedland; Michele Gallery; Chris Geist; Jean Geist; Bonnie Gellas; Richard Gertner; Amy Glass; Bea Gordon; Dan Gordon; Ken Gordon; Judy Goulding; Chris Graves; Adam Henner; Christal Henner; Lorin Henner; Marilu Henner; Melodie Henner; David Hennes; Paula Hennes; Sheila Hennes; Sophie Hennes; Larry Harold; Carl Hess; Mitchell Hofing; Steve Hofman; Bill Hohauser; Uday Ivatury; Terry Johnson; Sarah Jones; Allen Kahn; Mitch Kahn; Joel Kaplan; Dimi Karras; Maria Katinos; Stewart Kellerman; Harvey Kleinman; Claire Labine; Randy Ladenheim-Gil; Debbie Leitner; Marilyn Levin; Vicky Levy; Jared Lilienstein; Pattie Magee; Jack Mahoney; everyone at the Manhattan Bridge Club; Phil Martin; Chris McCann; Jeff McQuain; Julie Mears; Phil Mears; Carol Miller; Barbara Morrow; Honor Mosher; Phil Neel; Steve Nellisen; Craig Nelson; Millie North; Milt North; Charlie Nurse; Debbie Nye; Tom O'Brien; Pat O'Conner; Joanna Parker; Jeannie Perkins; Merrill Perlman; Joan Pirkle; Larry Prussin; Joe Rowley; Rose Reiter; Brian Rose; Lorraine Rose; Paul Rosenbaum; Carol Rostad; Tim Rostad; Leslie Rugg; Tom Rugg; Gary Saunders; Joan Saunders; Mike Saunders; Norm Saunders; Laura Schisgal; Cindy Shaha; Patricia Sheinwold; Kathy Smith; Kurtwood Smith; Susan Sherman Smith; Chris Soule; Kitty Srednicki; Karen Stoddard; Bill Stranger; Kat Stranger; Anne Swanson; Ed Swanson; Mike Szala; Jim Teuscher; Josephine Teuscher; Laura Tolkow; Carol Vellucci; Dan Vellucci; Hattie Washington; Ron Weinstock; Roy Welland; Dennis Whelan; Devin Whelan; Heide Whelan; Lara Whelan; Jon White; Ann Whitney; Carol Williams; Maggie Wittenburg; Karen Wooldridge; Maureen Wylie; Charlotte Zdrok; Vladimir Zdrok; and Debbie Zuckerberg.

The *Imponderables* books wouldn't be possible without the cooperation of experts in every subject from acting to zoology.

For this book, we contacted nearly 1,500 corporations, educational institutions, foundations, trade associations, and miscellaneous experts to find answers to our readers' Imponderables. Usually, there is nothing to gain for these sources other than the psychic benefit of sharing their knowledge. The following generous people, although not the only ones to supply help, gave us information that led directly to the solution of the Imponderables in this book:

Dr. Robert D. Altman, A & A Veterinary Hospital; American Academy of Dermatology; Dr. Duane Anderson, Central States Anthropological Society; Scott Anderson, Anheuser-Busch; Dan Arcy, Pennzoil Products; Dr. Harry Arnold; Sandi Atkinson.

Jim Ball, Dr Pepper/Seven-Up Companies; Michele Ball, National Audubon Society; Dr. Margaret Downie Banks, American Musical Instrument Society; Dr. Joseph Bark; J. P. Barnett, South Bend Replicas; Bausch & Lomb, Inc.; Ralph Beatty, Western/English Retailers of America; Barbara Begany, Dellwood Milk; Prof. George Bergman, University of California; Brian Bigley; Biff Bilstein, Neodata Services; Michelle Bing, United Fresh Fruit & Vegetable Association; Peter Black, American Water Resources Association; Harold Blake; Prof. Dee Boersma, University of Washington; Stephen Bomer, Automotive Battery Charger Manufacturers; G. Bruce Boyer; Frank Brennan, United States Postal Service; Dr. Donald Bruning, New York Zoological Park; Lloyd Brunkhorst, Brown Shoe Company; Robert Burnham; Trish Butler, Social Security Administration.

Jim Cannon; Roger W. Cappello, Clear Shield National; Thomas J. Carr, Motor Vehicle Manufacturers Association; Louis T. Cerny, American Railway Engineers Association; C. R. Cheney, Chrysler Motors; John Chuhran, Mercedes-Benz of North America; John Clark, Social Security Administration; Catherine Clay, Florida Department of Citrus; Prof. Richard Colwell, Council for Research in Music Education; Charlotte Connelly, Whitman's Chocolates; John Corbett, Clairol, Inc.; Capt. Kenneth L. Coskey, Navy Historical Foundation; Dr. Regis Courtemanche, C. W. Post; Brian Cudahy, Urban Mass Transportation

DAVID FELDMAN

Administration; Todd Culver, Cornell University Laboratory of Ornithology; Fred A. Curry.

Dr. Frank Davidoff, American College of Physicians; Bill Deane, Hall of Fame Museum; William Debuvitz; Roger De-Camp, National Food Processors Association; Richard Decker, International Conference of Symphony and Opera Musicians; Thomas H. Dent, Cat Fanciers Association; Robin Diamond, American Bus Association; Dr. Liberato DiDio, International Federation of Associations of Anatomists; Claire O'Neill Dillie; David DiPasquale, DiPasquale & Associates; Donna Ditmars, M&M/Mars; Sara Dornacker, United Airlines; Richard H. Dowhan, GTE Products Corporation; Ed Dunn, Cramer-Krasselt; Steven Duquette, National Cartoonists Society.

Mark Earley, Neodata Services; Charley Eckhardt; Morris Eckhouse, SABR; Prof. Gary Elmstrom, University of Florida; Dale Elrod, Jack K. Elrod Co.; Kathleen Etchpare, *Bird Talk*.

Raymond Falconer, SUNY at Albany; Michael Falkowitz, Nabisco Brands; Rob Farson, Neodata Services; Peter Fenimore, New York Aquarium; Stanley Fenvessy, Fenvessy Consulting; Karen Finkel, National School Transportation Association; Tim Fitzgerald; George Flower; Carol Frasier, Montana Historical Society; Eileen Frech, Thomas English Muffins; Don French, Radio Shack; Francis Frere, United States Mint.

Stan S. Garber, Selmer Co.; R. Bruce Gebhardt, North American Native Fishes Association; David A. Gibson, Eastman Kodak Company; Mark Gill, Columbia Pictures Entertainment; Mary Gillespie, Association of Home Appliances; Martin Gitten, Consolidated Edison; Dr. Paul Godfrey, Water Resources Research Center; William Goffi, Maxell Corporation of America; Tamara Goldman, *Food & Beverage Marketing;* Stanley Gordon, Federal Highway Administration; Robert Grayson, Grayson Associates; Dr. E. Wilson Griffin III, Jonesville Family Medical Center; David Guidry, Ushio America.

Susan Habacivch, DuPont; Fran Hackett, New York Aquarium; Brian Hannan, Urban Mass Transportation; Charles E. Hanson, Museum Association of the American Frontier; Joseph

Hanson, Hanson Publishing Group; Larry Hart, Talon Inc.; Sylvia Hauser, *Dog World;* Jeanette Hayhurst; Peter Heide, Association of Manufacturers of Confectionary and Chocolate; Prof. John Hertner, Kearney State College; Dr. James Riley Hill, Clemson University; Ruth Hill, Internal Revenue Service; Scott Hlavaty, Angelica Uniform Group; Dick Hofacker, AT&T Bell Laboratories; Beverly Holmes, Frito-Lay, Inc.; W. Ray Hyde.

Pete James, National Association of Pupil Transportation; Dr. Ben H. Jenkins; Alvin H. Johnson, American Musicological Society; Lloyd Johnson, SABR; Mark Johnson, Matsushita Electric Corporation of America; Dr. William P. Jollie, American Association of Anatomists; Chris Jones, Pepsico; George E. Jones, National Highway Institute; Ed Juge, Radio Shack.

Phil Katz, Beer Institute; Prof. George Kauffman, California State University, Fresno; William Kelly, Brockton Sole and Plastics; Rose Marie Kenny, Hammerhill Papers; Robert E. Kenyon, American Society of Magazine Editors; Dr. Wayne O. Kester, American Association of Equine Practitioners; Frank Kiley; Jula Kinnaird, National Pasta Association; Klousseau, New York Aquarium; Kevin Knopf, Department of the Treasury; Irving Smith Kogan, Champagne Association; Robert L. Krick, Federal Railroad Administration; Lucille Kubichek.

Dr. Eugene LaFond, International Association for Physical Sciences of the Ocean; Christine Lamar, Rhode Island Secretary of State's Office; Prof. Richard Landesman, University of Vermont; William L. Lang, Columbia-Wren; Fred Lanting; Michael Lauria; Carol Lawler, Land O' Lakes, Inc.; Thomas A. Lehmann, American Institute of Baking; Dr. Jay Lehr, Association of Ground Water Science and Engineering; Dr. Jerome Z. Litt; Richard Livingston, Airline Passengers Association.

David MacKenzie, University of Northern Colorado; Alan MacRobert, *Sky & Telescope;* Tom Mancini, U. S. Polychemical Corporation; Nancy Martin, Association of Field Ornithologists; John Matter, National Ballroom and Entertainment Association; Doug Matyka, Georgia-Pacific Corporation; Mauna Loa Macadamia Nut Company; Karen E. McAliley, United States Postal System; John T. McCabe, Master Brewers Association of the

DAVID FELDMAN

Americas; Diane McCulloch, Mount de Chantal Academy; Larry McFather, International Brotherhood of Locomotive Engineers; Jil McIntosh; Dr. Jim McKean, Iowa State University; Art McNally, National Football League; Mary Medin, "How to with Pete"; Jim Meyer, United States Postal System; Jerry Miles, American Baseball Coaches Association; H. Dale Millay, Shell Development Company; F. Kent Mitchel, Marketing Science Institute, Thomas Mock, Electronic Industries Association; Nevin B. Montgomery, National Frozen Food Association; Dr. David Moore, Virginia Tech University; Jeffrey Mora, Urban Mass Transportation Office; John M. Morse, Merriam-Webster, Inc.; Claude Mouton, Montreal Canadiens.

National Mapping Division, Department of Interior; Niagara Straw Company; Robert Nichter, Fulfillment Management Association; Robert Niddrie, Playtex Apparel.

Richard T. O'Connell, Chocolate Manufacturers of the USA; Bill O'Connor, Topps Chewing Gum; Doug Olander, International Game Fish Association; Karen Orme, *Footwear Forum.*

Dennis Patterson, Murray Bicycle Company; Bill Paul, *School Bus Fleet;* Roger Payne, Department of Interior; Roger S. Pinkham, Stevens Institute of Technology; John A. Pitcher, Hardwood Research Council; Joseph Pocius, National Turkey Federation; Prof. Anthony Potter, University of Hawaii at Manoa.

Dr. T. E. Reed, American Rabbit Breeders Association; Prof. Billy Rhodes, Poole Agricultural Center; Al Rickard, Snack Food Association; Robert S. Robe Jr., Scipio Society of Naval and Military History; Ronnie Robertson; Dr. Robert R. Rofen, Aquatic Research Institute; Barbara Rose, Continental Baking; Prof. Neal Rowell, University of South Alabama; Thomas Ruble; Tom and Leslie Rugg; Oscar Mayer Foods; Max Rumbaugh, SAE.

Dr. John Saidla, Cornell Feline Health Center; Kim Sakamoto, California Melon Research Board; Norman Savig, University of Northern Colorado; Dr. Charles Schaefer, University of California at Berkeley; Robert Schmidt, North American Native Fishes Association; Kassie Schwan; Dr. Samuel Selden; Vickie Sheer, Dance Educators of America; Bill Shoenleber, Edmund

Scientific Company; Carole Shulman, Professional Skaters' Guild of America; F. G. Walton Smith, International Oceanographic Foundation; Prof. Stephen Smulski, University of Massachusetts; Bruce V. Snow; Charles Spiegel; Cherie Spies, Continental Baking; Amy Steiner, American Association of State Highway and Traffic Officials; Bob Stewart, Association of American Railroads; David Stivers, Nabisco Brands; Lisa Stormer; Amurol Products Company; Peggy Sullivan, Music Educators National Conference; John J. Surrette, Rolls Battery Engineering; Barbara Sweeney, AT&T Library Network Archives.

Farook Taufiq, Prince Company; Dr. Kristin Thelander, University of Iowa; Susan Tildesley, Headwear Institute of America; Fran Toth, Cadillac Motor Car Division; Constance Townsend, U. S. Amateur Ballroom Dancers Association; Jim Trdinich, National League; Ray Tricarico, Playtex International.

Mary Ann Usrey, R. J. Reynolds Tobacco Company.

John Veltman; Carolyn Verweyst, Whirlpool Corporation.

A. R. Ward, Railroadians of America; Pat Weissman, American Association of Railroad Superintendents; Richard Williams; Bruce Wittmaier; Peter Wulff, *Home Lighting & Accessories*.

YKK Inc.

Caden Zollo, The Specialty Bulb Company, Inc.

Index

Magazines *(cont.)*
    subscription insert cards in,
        157–58
Mail
    CAR-RT SORT on envelopes,
        78–79
    undeliverable, 13–14
Mail-in refunds, vs. coupons,
    187–88
Marching, stepping off on left
    foot when, 172–73
Margarine sticks, length of, 42
Marshals' badges, shape of, 73–
    74
Meat, age and doneness
    preferences in, 230–31
Meat loaf, taste of in institutions,
    203
Menthol, coolness of, 192
Milk cartons, design of, 112, 243
Milk cases, warnings on, 43–44
Milk in refrigerators, coldness of,
    4–5
Mint, U.S., and shipment of coin
    sets, 32
Montreal Canadiens uniforms,
    165
Moon
    effect on lakes and ponds, 138–
        39
    official name of, 19–20
Mosquitoes
    biting and itching, 3–4
    daytime habits of, 77–78
    male vs. female eating habits
        of, 190
Movie actors and speed of
    speech, 203

Nabisco Shredded Wheat box,
    Niagara Falls on, 100–101
National Geographics, saving of,
    229–30

Newspapers, jumps in, 116–17
New York City and steam in
    streets, 16–17
Niagara Falls on Nabisco
    Shredded Wheat box, 100–
    101
911 as emergency telephone
    number, 145–46

Oceans
    salt in, 149–50
    vs. seas, 30–32
Oil, automobile, after oil change,
    184–85
Orange juice, price of fresh vs.
    frozen, 155–56
Oxygen masks, inflation of
    airline, 196–98

Paper mills, smell of, 96–98
Penguins, knees of, 160
Penmanship of doctors, 201
Pens, disappearance of, 222–23
Pepper, and salt, as table
    condiments, 201
Pepsi-Cola, and trademark
    location, 115–16
"Pi" as geometrical term, 80–81
Pigs and curly tails, 218–19
Pitcher's mound, location of, 181
Plug prongs
    holes at end of, 94–95
    three prongs vs. two prongs,
        191
Plum pudding, plums in, 49
Ponds
    effect of moon on, 138–39
    ice formations on, 82–83
    vs. lakes, 29–30
Postal Service, U.S., and
    undeliverable mail, 13–14
Potato chips
    green tinges in, 136–37

Trains *(cont.)*
   EXEMPT signs at railroad
      crossings, 118–19

United States Mint and shipment
   of coin sets, 32
UPS and shipment of coin sets,
   32
United States Postal Service
      (USPS)
   and CAR-RT SORT on
      envelopes, 78–79
   and undeliverable mail, 13–14

Videotape recorders
   "play" and "record" switches
      on, 23–24
   and storms, 180–81

Videotapes, rental, two-tone
   signals on, 144–45

Water, manufacture of, 107–8
Water faucets, bathroom vs.
   kitchen, 244
Watermelon seeds, white vs.
   black, 94
Water towers, height of, 91–93
White chocolate, vs. brown
   chocolate, 134–35
Windows, rear, of automobiles,
   143–44
Wines, and dryness, 141–42

YKK on zippers, 180

Zippers, YKK on, 180

Master Index of Imponderability

Following is a complete index of all ten Imponderables® books and *Who Put the Butter in Butterfly?* The bold number before the colon indicates the book title (see the Title Key below); the numbers that follow the colon are the page numbers. Simple as that.

Title Key

Airplanes (*cont.*)
"Qantas," spelling of, **8**:134–135
red and green lights on, **4**:152–153
seat belts, **8**:141–142
shoulder harnesses, **8**:141–142; **9**:296
U.S. flags on exterior, **8**:36–37
Alarm clocks, life before, **9**:70–74
Alcohol
in cough medicine, **3**:166
proof of, **2**:177
Algebra, X in, **9**:131–132
"All wool and a yard wide," origins of term, **11**:2
"Allemande," **11**:31
"Alligator" shirts, **9**:297–298
Alphabet, order of, **1**:193–198
Alphabet soup
distribution of letters, **3**:118–119
outside of U.S., **10**:73
Aluminum cans, crushability of, **7**:157–159
Aluminum foil
and heat to touch, **8**:145–146
on neck of champagne bottles, **4**:160–161
two sides of, **2**:102
Ambidexterity in lobsters, **6**:3–4
Ambulances, snake emblems on, **6**:144–145; **7**:239–240
American accents of foreign singers, **4**:125–126
American singles, Kraft, milk in, **1**:247–249
"Ampersand," **11**:86
Amputees, phantom limb sensations in, **1**:73–75
Anchors, submarines and, **4**:40–41

Angel food cake and position while cooling, **7**:43–44
Animal tamers and kitchen chairs, **7**:7–11
Ants
separation from colony, **6**:44–45
sidewalks and, **2**:37–38
"Apache dance," **11**:30
Apes, hair picking of, **3**:26–27
Appendix, function of, **5**:152–153
Apples, as gifts for teachers, **2**:238; **3**:218–220
Apples and pears, discoloration of, **4**:171
Apples in roasted pigs' mouths, **10**:274–275
April 15, as due date for taxes, **5**:26–29
Aquariums, fear of fish in, **4**:16–18
Arabic numbers, origins of, **3**:16–17
Architectural pencils, grades of, **7**:73
Area codes, middle digits of, **5**:68–69; **9**:287
Armies, marching patterns of, **8**:264
Armpits, shaving of, **2**:239; **3**:226–229; **6**:249
Army and Navy, Captain rank in, **3**:48–50
Art pencils, grades of, **7**:73
Aspirin
headaches and, **7**:100–102
safety cap on 100-count bottles of, **4**:62
Astrology, different dates for signs in, **4**:27–28
Astronauts and itching, **9**:208–216
"At loggerheads," **11**:104–105
Athletics, Oakland, and elephant insignia, **6**:14–15

Baskin-Robbins, cost of cones versus cups at, **1**:133–135
"Batfowling," **11**:1–2
Bathrooms
group visits by females to, **7**:183–192; **8**:237–238; **9**:277–278
ice in urinals of, **10**:232–234
in supermarkets, **6**:157
Bathtub drains, location of, **3**:159–160
Bathtubs, overflow mechanisms on, **2**:214–215
Bats, baseball, stripes on, **8**:104–106
Batteries
automobile, weight of, **5**:101–102
concrete floors and, **10**:232–234
drainage of, in cassette players, **10**:259–260
nine-volt, shape of, **6**:104; **7**:242–243
sizes of, **3**:116
volume and power loss in, **2**:76–77
"Battle royal," **11**:67
Bazooka Joe, eye patch of, **5**:121
Beacons, police car, colors on, **7**:135–137
"Bead," "Draw a," origins of term, **10**:168
Beaks versus bills, birds and, **10**:3–4
Beanbag packs in electronics boxes, **6**:201
Beans, green, "French" style, **10**:125–126
Beards on turkeys, **3**:99
"Bears [stock market]," **11**:106–107

"Beating around the bush," **11**:1–2
Beavers, dam building and, **10**:42–46
"Bedlam," **11**:69
Beds, mattresses, floral graphics on, **9**:1–2
Beef, red color of, **8**:160–161
Beeps before network news on radio, **1**:166–167
Beer
and plastic bottles, **7**:161–162
steins, lids of, **9**:95–96
temperature in Old West, **5**:17–18
twistoff bottle caps, merits of, **4**:145
Beetle, Volkswagen, elimination of, **2**:192–194
Bell bottoms, sailors and, **2**:84–85
Bells in movie theaters, **1**:88–89
Belly dancers, amplitude of, **5**:202–203; **6**:237–239
Belts, color of, in martial arts, **9**:119–123
Ben-Gay, creator of, **6**:46–47
"Berserk," **11**:68–69
Best Foods Mayonnaise, versus Hellmann's, **1**:211–214
Beverly Hills, "Beverly" in, **8**:16–17
"Beyond the pale," **11**:3
Bias on audiotape, **4**:153–154
Bibles, courtrooms and, **3**:39–41
Bicycles
clicking noises, **5**:10–11
crossbars on, **2**:90–91
tires, **2**:224–226
Bill posting at construction sites, **8**:185; **9**:268–270
Billboards, spinning blades on, **9**:61–64

MASTER INDEX OF IMPONDERABILITY

Cattle guards, effectiveness of, **3**:115

"Cattycorner," **11**:197

Cavities, dogs and, **10**:277–278

CDs, Tuesday release of, **10**:108–112

Ceiling fans
direction of blades, **9**:113–114
dust and, **9**:111–113; **10**:263–264

Ceilings of train stations, **8**:66–67

Celery in restaurant salads, **6**:218; **7**:207–209; **8**:249

Cement
laying of, **8**:258
versus concrete, **9**:295

Cemeteries
financial strategies of owners of, **2**:95–99
perpetual care and, **2**:221–222

Ceramic tiles in tunnels, **6**:135–136; **8**:257–258

Cereal
calorie count of, **1**:38–40
joining of flakes in bowl, **8**:115–119
Snap! and Rice Krispies, **8**:2–4

Chalk outlines of murder victims, **3**:11–12

Champagne
aluminum foil on neck of, **4**:160–161
name of, versus sparkling wine, **1**:232–234

Channel 1, lack of, on televisions, **4**:124; **7**:242; **10**:267–268

Chariots, Roman, flimsiness of, **10**:105–107

"Checkmate," **11**:133

Checks
approval in supermarkets, **7**:210–217; **8**:245–246

numbering scheme of, **4**:38–39; **5**:236–237
out-of-state acceptance of, **8**:120–121
white paper attachments, **6**:124–125; **7**:245

Checks, canceled
numbers on, **6**:123
returned, numerical ordering of, **6**:165–166
white paper attachments, **6**:124–125; **7**:245

Cheddar cheese, orange color of, **3**:27–28; **10**:275

Cheese
American, milk in Kraft, **1**:247–249
cheddar, orange color of, **3**:27–28; **10**:275
string, characteristics of, **3**:155
Swiss, holes in, **1**:192
Swiss, slice sizes of, **9**:142–146

Chef's hat, purpose of, **3**:66–67

Chewing gum
lasting flavor, **5**:195–196
water consumption and hardening of, **10**:236–237
wrapping of, **8**:111–112

Chewing motion in elderly people, **7**:79–80

Chianti and straw-covered bottles, **8**:33–35

Chicken
cooking time of, **1**:119–121
versus egg, **4**:128
white meat versus dark meat, **3**:53–54

"Chicken tetrazzini," **11**:153

Children, starving, and bloated stomachs, **7**:149–150

Children's reaction to gifts, **8**:184; **9**:234–237

MASTER INDEX OF IMPONDERABILITY

Chime signals on airlines, **7**:6–8
Chirping of crickets, at night, **10**:54–57
Chocolate
Easter bunnies, **2**:116
shapes of, **2**:24–25
white versus brown, **5**:134–135
wrapping of boxed, **8**:122–123
Chocolate milk, consistency of, **3**:122–123
"Chops," **11**:47
Chopsticks, origins of, **4**:12–13
"Chowderhead," **11**:72
Christmas card envelopes, bands around, **6**:203–204
Christmas tree lights
burnout of, **6**:65–66
lack of purple bulbs in, **6**:185–186; **9**:293; **10**:280
Cigar bands, function of, **4**:54–55
Cigarette butts, burning of, **5**:45
Cigarettes
grading, **6**:112
odor of first puff, **2**:238; **3**:223–226
spots on filters, **6**:112–113
Cigars, new fathers and, **3**:21–22
Cities, higher temperatures in, compared to outlying areas, **1**:168–169
Civil War, commemoration of, **3**:168–169
"Claptrap," **11**:73
Clasps, migration of necklace and bracelet, **7**:180; **8**:197–201; **9**:279–281
Cleansers, "industrial use" versus "household," **5**:64–65
Clef, treble, dots on, **10**:210–213
Clicking noise of turn signals, in automobiles, **6**:203
Climate, West Coast versus East coast, **4**:174–175

Clocks
clockwise movement of, **2**:150
grandfather, **4**:178
number 4 on, **2**:151–152
Roman numerals, **5**:237–238
school, backward clicking of minute hands in, **1**:178–179
versus watches, distinctions between, **4**:77–78
Clockwise, draining, south of the Equator, **4**:124–125
"Cloud nine," **11**:97–98
Clouds
disappearance of, **5**:154
location of, **3**:13
rain and darkness of, **2**:152
Clouds in tap water, **9**:126–127
"Cob/cobweb," **11**:20
Coca-Cola
2-liter bottles, **5**:243–244
taste of different size containers, **2**:157–159
Cockroaches
automobiles and, **7**:34; **8**:256–257; **9**:298
death position of, **3**:133–134; **8**:256
reaction to light, **6**:20–21
Coffee
bags in lavatories of airplanes, **4**:64–65
bags versus cans, **4**:146
electric drip versus electric perk in, **4**:35
restaurant versus home-brewed, **7**:181; **8**:221
Coffee, decaffeinated
leftover caffeine usage, **6**:195
orange pots in restaurants, **6**:67–69
Coffeemakers, automatic drip, cold water and, **4**:173

"Dukes," **11**:137
"Dumb [mute]," **11**:131–132
"Dumbbells," **11**:131–132
Dust, ceiling fans and, **10**:263–264

E
   as school grade, **3**:198; **4**:206–209
   on eye charts, **3**:9–10
"Eagle [golf score]," **11**:139–140
Earlobes, function of, **5**:87–88
"Earmark," **11**:46
Earrings, pirates and, **9**:43–45;
      10:272–273
Ears
   hairy, in old men, **2**:239;
      3:231–233; **5**:227–228
   popping in airplanes, **2**:130–132
   ringing, causes of, **2**:115–116
Earthworms as fish food, **3**:110–112
Easter
   chocolate bunnies and, **2**:116
   dates of, **4**:55–56
   ham consumption at, **1**:151–152
"Easy as pie," **11**:172
Eating, effect of sleep on, **6**:138–139
"Eavesdropper," **11**:109–110
Ebert, Roger, versus Gene Siskel,
      billing of, **1**:137–139
Egg, versus chicken, **4**:128
Egg whites and copper bowls,
      7:99
Eggs
   color of, **2**:189–190
   double-yolk, **3**:188–189
   hard-boiled, discoloration of,
      3:34

meaning of grading of, **4**:136–137
   sizes of, **2**:186–188
"Eggs Benedict," **11**:154
"Eighty-six," origins of term,
      10:265–266; **11**:101–102
Elbow macaroni, shape of, **4**:28
Elderly men
   pants height and, **2**:171–172
   shortness of, **2**:239; **3**:229–231;
      6:250
Elections, U.S.
   timing of, **6**:41; **8**:260–261;
      9:291–292
   Tuesdays and, **1**:52–54; **3**:239
Electric can openers, sharpness of
      blades on, **6**:176–177
Electric drip versus electric perk,
      in coffee, **4**:35
Electric perk versus electric drip,
      in coffee, **4**:35
Electric plug prongs
   holes at end of, **5**:94–95
   three prongs versus two prongs,
      5:191
Electricity, AC versus DC, **2**:21–22
Electricity, static, variability in
      amounts of, **4**:105–106
Elephants
   disposal of remains of, **6**:196–197
   jumping ability of, **10**:27–29
   Oakland A's uniforms, **6**:14–15
   rocking in zoos, **8**:26–27;
      10:279
Elevator doors
   changing directions, **8**:169–170
   holes, **8**:170–171
Elevators
   overloading of, **5**:239
   passenger capacity in, **1**:23–24
"Eleventh hour," **11**:99–100

Flags, U.S.
    half-mast, **10**:36–38
    on airplanes, **8**:36–37
"Flak," **11**:74
"Flammable," versus "inflamma-
    ble," **2**:207–208
"Flash in the pan," **11**:74
Flashlights, police and grips of,
    10:30–32
Flat toothpicks, purpose of,
    1:224–225
"Flea market," **11**:24
Flies
    landing patterns of, **2**:220
    winter and, **2**:133–135
Flintstones, Barney Rubble's pro-
    fession on, **7**:173–174
Flintstones multivitamins, Betty
    Rubble and, **6**:4–5; **9**:285–
    286
Floaters in eyes, **3**:37
Floral graphics on mattresses,
    9:1–2
Florida, cost of auto rentals in,
    4:24–25
"Flotsam," versus "jetsam," **2**:60–
    61
Flour, bleached, **3**:63–64
Flour bugs, provenance of, **4**:89–
    90
Flour tortillas, size of, versus corn,
    10:142–145
Flu, body aches and, **3**:104–105
Fluorescent lights and plinking
    noises, **5**:47
Flush handles, toilets and, **2**:195–
    196
Flushes, loud, toilets in public
    restrooms and, **4**:187
Fly swatters, holes in, **4**:31–
    32
FM radio, odd frequency numbers
    of, **10**:59–60

Food cravings in pregnant women,
    10:183–185
Food labels
    "FD&C" on label, **4**:163
    lack of manufacturer street ad-
        dresses, **4**:85
Football
    barefoot kickers in, **4**:190–191
    college, redshirting in, **7**:46–48
    distribution of game balls,
        2:44
    goalposts, tearing down of,
        7:181; **8**:213–217
    measurement of first-down
        yardage, **5**:128–129
    origins of "hut" in, **9**:294–295
    Pittsburgh Steelers' helmet em-
        blems, **7**:67–68
    shape of, **4**:79–81
    sideline population in, **10**:51–
        53
    two-minute warning and,
        10:150–151
    yardage of kickers in, **3**:124–
        125
"Fore," origins of golf expression,
    2:34
Forewords in books, versus intro-
    ductions and prefaces, **1**:72–
    73
Forks, switching hands to use,
    3:198
"Fortnight," **11**:194
Fraternities, Greek names of,
    10:94–98
"Freebooter," **11**:110
Freezer compartments, location
    of, in refrigerators, **2**:230–
    231
Freezers
    ice trays in, **10**:92–93
    lights in, **10**:82–85
French dry cleaning, **3**:164–165

French horns, design of, **5**:110–111

"French" bread versus "Italian," **7**:165–166; **8**:261–262

"French" style green beans, **10**:125–126

Frogs
eye closure when swallowing, **6**:115–116
warts and, **10**:121–123

Frogs of violins, white dots on, **4**:164–165; **9**:291

Frostbite, penguin feet and, **1**:217–218

Fruitcake, alleged popularity of, **4**:197; **5**:205–209; **6**:253; **7**:226–227; **9**:274–276

"Fry," **11**:179

Fuel gauges in automobiles, **6**:273

"Fullback," **11**:138

Full-service versus self-service, pricing of, at gas stations, **1**:203–209

Funeral homes, size of, **8**:152–155

Funerals
burials without shoes, **7**:53–54
depth of graves, **7**:14–15
head position of, in caskets, **6**:8–9
orientation of deceased, **7**:106
perpetual care and, **2**:221–222
small cemeteries and, **2**:95–99

"Funk," **11**:75

"G.I.," **11**:52

Gagging, hairs in mouth and, **7**:76–77

Gallons and quarts, American versus British, **4**:114–115

Game balls, distribution of football, **2**:44

Gasoline, pricing of, in tenths of a cent, **2**:197–198

Gasoline gauges, automobile, **3**:5–6

Gasoline pumps
rubber sleeves on, **6**:197
shut off of, **3**:125

Gasoline stations, full-service versus self-service, pricing of, **1**:203–209

Gasoline, unleaded, cost of, **1**:121–122

Gauge, width of railroad, **3**:157–159

Gauges, fuel, in automobiles, **3**:5–6; **6**:273

Geese, honking during migration, **7**:108

Gelatin, fruit in, **3**:149–150

"Gerrymander," **11**:111–112

"Get the sack," **11**:75

"Get your goat," **11**:21

"Getting down to brass tacks," **11**:6

Gifts, children and, **8**:184

Girdles and fat displacement, **5**:75–76

Glass, broken, **4**:168

Glasses, drinking
squeakiness of, **9**:205–207
"sweating" of, **9**:124–125; **10**:261

Glasses, wine, types of, **7**:123–125

Glitter, sidewalks and, **10**:61–62

Glow-in-the-dark items, green color of, **9**:139–141

Glue
stickiness in the bottle, **7**:18
"Super" versus conventional, **6**:145–146
"Super," and Teflon, **7**:128–129
taste of postage stamp, **2**:182

Gum, chewing
  water consumption and hardening of, **10**:236–237
  wrappers of, **8**:111–112
"Gunny sacks," **11**:195
"Guy," **11**:151–152

"Habit [riding costume]," **11**:123
"Hackles," **11**:6–7
Hail, measurement of, **5**:203; **6**:239–240; **7**:234–235; **8**:236–237
Hair
  blue, and older women, **2**:117–118
  growth of, after death, **4**:163–164
  length of, in older women, **7**:179; **8**:192–197
  mole, color of, **8**:167–169
  parting, left versus right, **1**:116
Hair color, darkening of, in babies, **10**:209
Hair spray, unscented, smell of, **2**:184
Hairbrushes, length of handles on, **7**:38–39
Hairs in mouth, gagging on, **7**:76–77
Hairy ears in older men, **2**:239; **3**:231–233; **5**:227–228
Half dollars, vending machines and, **3**:54–56
"Halfback," **11**:138
Half-mast, flags at, **10**:36–38
Half-moon versus quarter moon, **7**:72–73
Half-numbers in street addresses, **8**:253
Halibut, coloring of, **3**:95–96
Halloween, Jack-o'-lanterns and, **4**:180–181

Halogen lightbulbs, touching of, **5**:164
Ham
  checkerboard pattern atop, **7**:66–67
  color of, when cooked, **7**:15–16
  Easter and consumption of, **1**:151–152
"Ham [actor]," **11**:170–171
Hamburger buns, bottoms of, **2**:32–34
"Hamburger," origins of term, **4**:125
"Hamfatter," **11**:170–171
Hand dryers in bathrooms, **8**:174–176; **10**:266–267
Hand positions in old photographs, **7**:24–26
Handles versus knobs, on doors, **7**:148–149
Handwriting, teaching of cursive versus printing, **7**:34–37
"Hansom cab," **11**:63
Happy endings, crying and, **1**:79–80
Hard hats
  backward positioning of, in ironworkers, **4**:94
  exterminators and, **2**:51
Hard-boiled eggs, discoloration of, **3**:34
Hat tricks, in hockey, **2**:165–166
Hats
  cowboy, dents on, **5**:6; **6**:274; **7**:249–250
  declining popularity, **5**:202; **6**:227–231; **7**:233; **7**:249
  holes in sides of, **5**:126
  numbering system for sizes, **4**:110
Haystacks, shape of, **6**:47–48; **8**:265–266
"Hazard [dice game]," **11**:134

"Head [bathroom]," **11**:48
"Head honcho," **11**:39
Head injuries, "seeing stars" and, **10**:156–158
Head lice, kids and, **10**:225–227
Headaches and aspirin, **7**:100–102
Headbands on books, **7**:126–127
Headlamps, shutoff of automobile, **7**:92–93
"Heart on his sleeve," **11**:128
Hearts, shape of, idealized versus real, **4**:199; **5**:220–221; **6**:260; **7**:229–230; **8**:234; **9**:234
Heat and effect on sleep, **6**:137–138
"Hector," **11**:155
"Heebie jeebies," **11**:40
Height
    clearance signs on highways, **8**:156–158
    of elderly, **6**:250
    restrictions on fences, **2**:28–30
    voice pitch and, **2**:70
Heinz ketchup labels, **8**:150–151
Helicopters, noise of, **8**:164–166
Helium and effect on voice, **5**:108–109
Hellmann's Mayonnaise, versus Best Foods, **1**:211–214
"Hem and haw," **11**:195–196
"Hep," **11**:52–53
Hermit crabs, bathroom habits of, **7**:74–75
Hernia exams and "Turn your head and cough," **5**:114–115
"Heroin," **11**:77
"High bias," versus "low bias," on audio tape, **4**:153–154
"High jinks," **11**:41
High-altitude tennis balls, **8**:80
"Highball," **11**:175–176

Highways
    clumping of traffic, **4**:165–167; **7**:247
    curves on, **7**:121–122
    interstate, numbering system, **4**:66–67
    traffic jams, clearing of, **1**:25–26
    weigh stations, **4**:193–194
"Hillbilly," **11**:148
Hills, versus mountains, **3**:97–98; **8**:252
"Hip," **11**:52–53
"Hobnob," **11**:40
"Hobson's choice," **11**:155–156
Hockey
    banging of sticks by goalies, **7**:116–117
    hat trick, **2**:165–166
    Montreal Canadiens uniforms, **5**:165; **7**:242
    Wayne Gretzky's uniform, **2**:18; **10**:279
"Hold a candle," **11**:71
"Holding the bag," **11**:75
Holes
    in barrels of cheap pens, **4**:111
    in elevator doors, **8**:170–171
    in fly swatters, **4**:31–32
    in ice cream sandwiches, **8**:128
    in needles and syringes, **10**:57–59
    in pasta, **4**:28
    in saltines, **8**:129
    in thimbles, **10**:63–64
    in wing-tip shoes, **8**:44
    on bottom of soda bottles, **6**:187–188
    recycling of, in loose-leaf paper, **7**:105–106
    refilling of dirt, **7**:48–49
"Holland," versus "Netherlands," **2**:65–66

Home plate, shape of, in baseball,
5:131
"Honcho," **11**:39
Honey, spoilage of, **4**:177–178
Honey roasted peanuts, banning
of, on airlines, **4**:13–14
Honking in geese during migration, **7**:108
"Honky," **11**:77
"Hoodwink," **11**:121
"Hoosiers," **11**:148–149
"Horsefeathers," **11**:40
Horses
measurement of heights of,
5:60–61
posture in open fields, **3**:104;
5:241
shoes, **3**:156
sleeping posture, **2**:212
vomiting, **6**:111–112; **7**:248
Hospital gowns, back ties on,
5:132–134
Hospitals and guidelines for medical conditions, **4**:75–76
Hot dog buns
number of, in package, **2**:232–
235
slicing of, **5**:161
Hot dogs, skins of, **5**:54
Hot water, noise in pipes of,
2:199–200
Hotels
amenities, spread of, **6**:118–121
number of towels in rooms,
4:56–57
plastic circles on walls of,
3:117
toilet paper folding in bathrooms of, **3**:4
"Hotsy totsy," **11**:40
Houses, settling in, **6**:32–34
"Hue and cry," **11**:112
"Humble pie," **11**:169

Humidity, relative, during rain,
1:225–226
Humming, power lines and,
10:259
Hurricane, trees and, **3**:68–69
"Hurricanes" as University of Miami nickname, **8**:171–172
"Hut," origins of football term,
6:40; **9**:294–295
Hydrants, fire, freezing water in,
10:11
Hypnotists, stage, techniques of,
1:180–191

"I [capitalization of]," **11**:55
"I could care less," **11**:78
"I" before "e," in spelling, **6**:219;
7:209; **8**:240–245
Ice
fizziness of soda, **9**:24–25
formation on top of lakes and
ponds, **5**:82–83
holes and dimples in, **9**:147–148
in urinals, **10**:232–234
Ice cream
black specks in, **8**:132–133
cost of cones versus cups,
1:133–135
pistachio, color of, **7**:12–13
thirstiness, **5**:202; **6**:236–237
Ice cream and soda, fizziness of,
9:27
Ice cream sandwiches, holes in,
8:128
Ice cubes
cloudy versus clear, **3**:106–107;
5:242
shape of, in home freezers,
5:103–104
Ice rinks, temperature of resurfacing water in, **10**:196–198
Ice skating, awful music in,
1:102–105

Ice trays in freezers, location of, **10**:92–93

Icy roads, use of sand and salt on, **2**:12–13

Ignitions
automobile, and headlamp shutoff, **7**:92–93
key release button on, **5**:169

Imperial gallon, versus American gallon, **6**:16–17

"In like Flynn," **11**:157

"In the nick of time," **11**:158

Index fingers and "Tsk-Tsk," stroking of, **4**:198; **5**:209–210

"Indian corn," **11**:146

"Indian pudding," **11**:146

"Indian summer," **11**:146

Indianapolis 500, milk consumption by victors in, **8**:130–131

"Inflammable," versus "flammable," **2**:207–208

Ink
color of, in ditto masters, **6**:133–134
. newspaper, and recycling, **7**:139–140

Insects
attraction to ultraviolet, **8**:158–159
aversion to yellow, **8**:158–159
flight patterns of, **7**:163–164
in flour and fruit, **4**:89–90
See also specific types

Insufficient postage, USPS procedures for, **4**:149–151

Interstate highways, numbering system of, **4**:66–67

Introductions in books, versus forewords and prefaces, **1**:72–73

Irish names, "O'" in, **8**:135–136

Irons, permanent press settings on, **3**:186–187

Ironworkers, backwards hard hat wearing of, **4**:94

Irregular sheets, proliferation of, **1**:145–147

IRS and due date of taxes, **5**:26–29

IRS tax forms
disposal of, **8**:143–144
numbering scheme of, **4**:9–10

"Italian" bread, versus "French," **7**:165–166; **8**:261–262

Itching, reasons for, **1**:172–173

Ivory soap, purity of, **2**:46–47

"J" Street, Washington, D.C., and, **2**:71

"Jack [playing card]," **11**:135

Jack Daniel's and "Old No. 7," **8**:144–145

"Jack," "John" versus, **2**:43

Jack-o'-lanterns, Halloween and, **4**:180–181

Jams, contents of, **6**:140–141

Japanese
baseball uniforms, **10**:207–208
boxes, yellow color of, **7**:130–131
flags, red beams and, **10**:151–155

Jars, food, refrigeration of opened, **6**:171–172

"Jaywalking," **11**:22–23

Jeans
blue, orange thread and, **9**:74
Levi's, colored tabs on, **6**:59–61
origin of "501" name, **6**:61
sand in pockets of new, **7**:152

"Jeans [pants]," **11**:124

"Jeep," **11**:61

Jellies, contents of, **6**:140–141

Jellies, grape, color of, **7**:142–143

Jell-O, fruit in, **3**:149–150

Jeopardy, difficulty of Daily Doubles in, **1**:33–35

"Jerkers," **11**:176

Jet lag, birds and, **3**:33–34

"Jetsam," versus "flotsam," **2**:60–61

"Jig is up," **11**:7

Jigsaw puzzles, fitting pieces of, **9**:3–4

Jimmies, origins of, **10**:165–168

"Jink," **11**:41

"John," versus "Jack," **2**:43

Johnson, Andrew, and 1864 election, **8**:85–87

"Joshing," **11**:158–159

Judges and black robes, **6**:190–192

Judo belts, colors of, **9**:119–123

"Juggernaut," **11**:63–64

Juicy Fruit gum, flavors in, **1**:71; **3**:242

"K rations," **11**:54–55

"K" as strikeout in baseball scoring, **5**:52–53

Kangaroos, pouch cleaning of, **4**:144–145

Karate belts, colors of, **9**:119–123

"Keeping up with the Joneses," **11**:159–160

Ken, hair of, versus Barbie's, **7**:4–5; **8**:259–260

Ketchup, Heinz, labels of, **8**:150–151

Ketchup bottles
  narrow necks of, **2**:44–45
  neck bands on, **5**:242
  restaurants mating of, **3**:200

"Ketchup," **11**:177

"Kettle of fish," **11**:178

Keys
  automobile, door and ignition, **3**:141–142
  piano, number of, **10**:7–9
  teeth direction, **8**:59–60
  to cities, **3**:99

"Kidnapping," **11**:113–114

Kids versus adult goats, **7**:64–65

Kilts, Scotsmen and, **7**:109–110

Kissing
  eye closure during, **7**:179; **8**:186–191
  leg kicking by females, **6**:218; **7**:196–197; **9**:278; **9**:299–300

"Kit cat club," **11**:38–39

"Kit," "caboodle" and, **2**:15–17

"Kittycorner," **11**:197

Kiwifruit in gelatin, **3**:149–150

Kneading and bread, **3**:144–145

Knee-jerk reflex in humans, **8**:255–256

Knives, dinner, rounded edges of, **1**:231–232

Knives, serrated, lack of, in place settings, **4**:109–110

Knobs versus handles, on doors, **7**:148–149

"Knock on wood," **11**:4–5

"Knuckle down," **11**:9–10

"Knuckle under," **11**:9

Knuckles, wrinkles on, **5**:182–183

Kodak, origins of name, **5**:169–170; **9**:290

Kool-Aid and metal containers, **8**:51

Kraft American cheese, milk in, **1**:247–249

"L.S.," meaning of, in contracts, **1**:165

Label warnings, mattress, **2**:1–2

Labels on underwear, location of, **4**:4–5

Labels, food, lack of manufacturer street addresses on, **4**:85

"Ladybug," **11**:23

Ladybugs, spots on, **7**:39–40

Lakes
   effect of moon on, **5**:138–139
   fish returning to dried, **3**:15–16; **10**:256
   ice formations on, **5**:82–83
   versus ponds, differences between, **5**:29–30; **7**:241
   versus ponds, water level of, **9**:85–86
   wind variations, **4**:156–157

"Lame duck," **11**:24–25

Lane reflectors, fastening of, **5**:98–99

Large-type books, size of, **5**:135

Laryngitis, dogs, barking, and, **2**:53–54

Lasagna, crimped edges of, **5**:61

"Last ditch," **11**:10

"Last straw," **11**:8–9

Laughing hyenas, laughter in, **8**:1–2

Lawn ornaments, plastic deer as, **8**:185

Lawns, reasons for, **2**:47–50

"Lawyer," **11**:103–104

"Lb. [pound]," **11**:56

Leader, film, **2**:9

"Leap year," **11**:193

Leather, high cost of, **8**:21–23

Ledges in buildings, purpose of, **8**:18–20

Left hands, placement of wrist-watches on, **4**:134–135; **6**:271

"Left wing," **11**:116

Left-handed string players, in orchestras, **9**:108–109; **10**:276–277

Leg kicking by women while

kissing, **6**:218; **7**:196–197; **9**:278; **9**:299–300

Legal-size paper, origins of, **3**:197

"Legitimate" theater, origins of term, **10**:5–7

"Let the cat out of the bag," **11**:25

Letters
   business, format of, **7**:180; **8**:201–204
   compensation for, between countries, **4**:5–6

Letters in alphabet soup, distribution of, **3**:118–119

Levi's jeans
   colored tabs, **6**:59–61
   origin of "501" name, **6**:61

Liberal arts, origins of, **5**:70–73

Lice, head, kids and, **10**:225–227

License plates and prisoners, **8**:137–139; **10**:268–269

License plates on trucks, absence of, **3**:98; **10**:270–271

"Licking his chops," **11**:47

Licorice, ridges on, **9**:188–189

Life Savers, wintergreen, sparkling of, when bitten, **1**:157–158

Lightbulbs
   air in, **6**:199–200
   fluorescent, stroking of, **3**:131
   halogen, **5**:164
   high cost of 25-watt variety, **5**:91
   in traffic signals, **3**:31–32
   loosening of, **3**:93–94
   noise when shaking, **5**:167
   plinking by fluorescent, **5**:47
   three-way, burnout, **2**:104
   three-way, functioning of, **2**:105
   wattage sizes of, **1**:255–256

Light switches, height and location of, **4**:183–184

"Make no bones about it," **11**:49

Mall, shopping, entrances, doors at, **6**:180–181

Mandarin oranges, peeling of, **8**:106–107

Manhole covers, round shape of, **3**:191

Marching, stepping off on left foot when, **5**:172–173; **9**:293–294

Marching bands, formations of, **8**:107–108

Margarine, versus butter, in restaurants, **1**:32–33

Margarine sticks, length of, **5**:42

Marmalades, contents of, **6**:140–141

Marshals' badges, shape of, **5**:73–74

Marshmallows, invention of, **8**:99–100

Martial arts, sniffing and, **10**:256–258

Martinizing, One Hour, **3**:28–29

Mascara, mouth opening during application of, **1**:257–260

Matchbooks, location of staples on, **6**:173–174

Matches, color of paper, **6**:174–175

Math, school requirement of, **8**:184; **9**:254–261

Mattress tags, warning labels on, **2**:1–2

Mattresses, floral graphics on, **9**:1–2

Maximum occupancy in public rooms, **10**:158–160

Mayonnaise, Best Foods versus Hellmann's, **1**:211–214

Mayors, keys to cities and, **3**:99

McDonald's
 Grimace, identity of, **7**:173
 "over 95 billion served" signs, **7**:171
 sandwich wrapping techniques, **7**:172
 straw size, **7**:171–172

Measurements
 acre, **2**:89
 meter, **2**:200–202

Measuring spoons, inaccuracy of, **1**:106–107

Meat
 children's doneness preferences, **5**:230–231; **6**:252–253; **9**:273–274
 national branding, **1**:227–231; **9**:287
 red color of, **8**:160–161

Meat loaf, taste in institutions, **5**:203; **6**:243; **7**:235–236

Medals, location of on military uniforms, **2**:223–224

Medical conditions, in hospitals, guidelines for, **4**:75–76

Medicine bottles, cotton in, **3**:89–90

Memorial Day, Civil War and, **3**:168–169

Men
 dancing ability of, **6**:218; **7**:199–202; **8**:239–240
 feelings of coldness, **6**:218; **7**:198–199
 remote controls and, **6**:217; **7**:193–196

Menstruation, synchronization of, in women, **4**:100–102

Menthol, coolness of, **5**:192

Meter, origins of, **2**:200–202

Miami, University of
 football helmets, **8**:171–172
 "Hurricanes" nickname, **8**:171–172

Mickey Mouse, four fingers of, **3**:32; **6**:271

Microphones, press conferences and, **2**:11–12

Migration of birds, **9**:91–94

Mile, length of, origins of, **1**:241–242

Military salutes, origins of, **3**:147–149

Milk
  as sleep-inducer, **7**:17
  fat content in lowfat, **7**:60–61
  in refrigerators, coldness of, **5**:4–5
  Indianapolis 500, **8**:130–131
  national brands, **1**:227–231; **9**:287
  plastic milk containers, **7**:61; **10**:262–263
  single serving cartons of, **7**:137–138
  skim versus nonfat, **7**:59
  skin on, when heated, **6**:58

Milk cartons
  design of, **5**:112; **9**:289–290
  difficulty in opening and closing of, **1**:243–246; **5**:243

Milk cases, warnings on, **5**:43–44

Milk Duds, shape of, **8**:81–82

Millimeters, as measurement unit for film, **1**:44

"Mind your P's and Q's," **11**:88–89

Mint flavoring on toothpicks, **4**:153

Mint, U.S., and shipment of coin sets, **5**:32

Minting of new coins, timing of, **3**:128

Mirrors in bars, **10**:14–17

Mirrors, rear-view, **4**:185–186

Mistletoe, kissing under, origins of, **4**:106–107

Mobile homes, tires atop, in trailer parks, **6**:163–164

Mole hair, color of, **8**:167–169

Money, U.S.
  color of, **3**:83–84
  stars on, **3**:180–182

Monitors, computer, shape of, **6**:129–131

Monkeys, hair picking of, **3**:26–27

Monopoly, playing tokens in, **10**:21–23

Montreal Canadiens, uniforms of, **5**:165; **7**:242

Moon
  apparent size of, at horizon, **2**:202–204
  effect on lakes and ponds, **5**:138–139
  official name, **5**:19–20
  quarter-, vs. half-, **7**:72–73

Moons on outhouse doors, **4**:126

Mosquitoes
  biting and itching, **5**:3–4
  biting preferences, **8**:177–179; **10**:278
  daytime habits, **5**:77–78
  male versus female eating habits, **5**:190

Moths, reaction to light of, **6**:21–23

Mottoes on sundials, **6**:54–56

Mountains
  falling hot air at, **9**:149–151
  versus hills, **3**:97–98; **8**:252

Movie actors and speed of speech, **5**:203; **6**:241–243

Movie theaters
  bells in, **1**:88–89
  in-house popcorn popping, **1**:45–50

Movies, Roman numerals in copyright dates in, **1**:214–216

"Mrs.," **11**:57

MSG, Chinese restaurants and, **2**:168–171

"Mugwump," **11**:119

Muppets, left-handedness of, **7**:111–113

Murder scenes, chalk outlines at, **3**:11–12

Musketeers, Three, lack of muskets of, **7**:29–30

Mustaches, policemen and, **6**:219; **7**:218–220; **8**:246–247; **9**:278

"Muumuu," **11**:125

"Mystery 7," in *$25,000 Pyramid*, **1**:192

Nabisco Saltine packages, red tear strip on, **1**:147–149

Nabisco Shredded Wheat box, Niagara Falls on, **5**:100–101

Nail polish and fingernail yellowing, **7**:129–130

"Namby pamby," **11**:79

National Geographics, saving issues of, **3**:199; **5**:229–230; **7**:224

Navy and Army, Captain rank in, **3**:48–50

Necklaces and clasp migration, **7**:180; **8**:197–201; **9**:279–281

Neckties
direction of stripes on, **6**:86–87
origins of, **4**:127; **8**:264–265
taper of, **6**:84–85

Nectarines, canned, lack of, **4**:59–60; **9**:287–288

Needles, holes in, of syringes, **10**:57–59

Neptune's moon, Triton, orbit pattern of, **4**:117–118

Nerdiness and eyeglasses, **7**:180

"Netherlands," versus "Holland," **2**:65–66

New York City and steam in streets, **5**:16–17

"New York" steaks, origins of, **7**:155–156; **8**:252

New Zealand, versus "Old Zealand," **4**:21–22

Newspapers
ink and recycling of, **7**:139–140
ink smudges on, **2**:209–212
jumps in, **5**:116–117
tearing of, **2**:64
window cleaning and, **10**:33–36
yellowing of, **8**:51–52

Niagara Falls on Nabisco Shredded Wheat box, **5**:100–101

"Nick of time," **11**:158

Nickels, smooth edges of, **1**:40–41

"Nickname," **11**:163

Nightclubs, lateness of bands in, **8**:184; **9**:248–254

"Nine-day wonder," **11**:98

Nine-volt batteries, shape of, **6**:104; **7**:242–243

Nipples, purpose of, in men, **4**:126; **6**:275

"No bones about it," **11**:49

"No Outlet" signs, versus "Dead End" signs, **4**:93

Noise, traffic, U.S. versus foreign countries, **4**:198

North Carolina, University of, and Tar Heels, **8**:76–77

North Pole
directions at, **10**:243
telling time at, **10**:241–243

Nose rings and bulls, **10**:147–148

Noses
clogged nostrils and, **3**:20–21
runny, in cold weather, **10**:146–147
runny, kids versus adults, **9**:89–90
wet, in dogs, **4**:70–73

Nostrils, clogged, **3**:20–21

Notches on bottom of shampoo bottles, **10**:29–30

Notre Dame fighting Irish, **10**:115–117

NPR radio stations, low frequency numbers of, **10**:181–183

Numbers, Arabic, origins of, **3**:16–17

Nutrition labels, statement of fats on, **6**:142–143

Nuts
Brazil, in assortments, **7**:145–147
Macadamia shells, **8**:262
peanuts in plain M&M's, **7**:239
peanuts, and growth in pairs, **7**:34

"O'" in Irish names, **8**:135–136

Oakland A's, elephant on uniforms of, **6**:14–15

Oboes, use of as pitch providers, in orchestras, **4**:26–27

Occupancy, maximum, in public rooms, **10**:158–160

Oceans
boundaries between, **10**:74–76
color of, **2**:213
salt in, **5**:149–150
versus seas, **5**:30–32

Octopus throwing, Detroit Red Wings and, **9**:183–186

"Off the schneider," **11**:136

Oh Henry!, origins of name of, **8**:83–84

Oil
automotive, after oil change, **5**:184–185; **7**:240–241
automotive, grades of, **3**:182–183

"Okay," thumbs-up gesture as, **1**:209–210

Oktoberfest, September celebration of, **9**:156–157

Old men
hairy ears and, **2**:239; **3**:231–233; **5**:227–228
pants height and, **2**:171–172; **6**:274

"Old No. 7" and Jack Daniel's, **8**:144–145

"Old Zealand," versus New Zealand, **4**:21–22

Olive Oil, virgin, **3**:174–175

Olives, green and black, containers of, **1**:123–127

"On pointe" and ballet, **8**:69–72

"On tenterhooks," **11**:10–11

"On the Q.T.," **11**:59

"Once in a blue moon," **11**:12

"One fell swoop," **11**:197

One Hour Martinizing, **3**:28–29

One-hour photo processing, length of black-and-white film and, **4**:39

Onions and crying, **9**:169–170

Orange coffee pots, in restaurants, **6**:67–69

Orange juice
price of fresh versus frozen, **5**:155–156
taste of, with toothpaste, **10**:244–246

Orange thread in blue jeans, **9**:74

Oranges, extra wedges of, **4**:175–176

Oranges, mandarin, peeling of, **8**:106–107

"Oreo," origins of name, **2**:173–174

Outhouse doors, moons on, **4**:126

Ovens, thermometers in, **10**:85–87

Overflow mechanism, kitchen sinks and, **2**:214–215

Oxygen in tropical fish tanks,
  7:84–85
Oxygen masks, inflation of airline,
  5:196–198

"P.U.," origins of term, 10:26–27
"Pacific Ocean," 11:149
Page numbers on magazines,
  4:14–15
Pagination in books, 1:141–144
Pain, effect of warmth upon,
  3:134–135
Paint, homes and white, 2:100–
  102
Paint, red, on coins, 7:117
Painters and white uniforms,
  6:17–19
Palms, sunburn on, 8:63–64
Pandas, double names of, 8:172–
  173
Pants, height of old men's,
  2:171–172; 6:274
"Pantywaist," 11:81
"Pap test," 11:49
Paper
  legal size, origins of, 3:197
  recycling of holes in loose-leaf,
    7:105–106
Paper cups, shape of, 9:289
Paper cuts, pain and, 2:103–104
Paper mills, smell of, 5:96–98
Paper sacks
  jagged edges on, 6:117–118
  names on, 2:166–167
Paper towel dispensers, "emer-
    gency feed" on, 8:149–150
Paperback books, staining of,
  2:93–94
Papers, yellowing of, 8:51–52
"Par [golf course]," 11:139–140
"Par Avion" on air mail postage,
  8:39
"Pardon my French," 11:150

Parking lots, sea gulls at, 10:254–
  256
Parking meters, yellow "violation"
    flags and, 4:42–43
"Parkway," 11:65
Parkways, parking on, versus
    driveways, 4:123
Parrots and head bobbing, 8:23–
  24
Parting of hair, left versus right,
  1:116
Partly cloudy, versus partly sunny,
  1:21–22
Partly sunny, versus partly cloudy,
  1:21–22
"Pass the buck," 11:107
Pasta
  boxes, numbers on, 4:107
  foaming when boiling, 7:78
  holes in, 4:28
Pay phones, collection of money
    from, 1:107–108
Pay toilets, disappearance of,
  2:25–26
PBX systems, 3:75–76
"Pea jacket," 11:124
Peaches
  canned, and pear juice, 8:46–
    47
  fuzziness of, 4:58–59
Peanut butter, stickiness of,
  10:204–207
Peanuts
  allergies to, 7:239
  growth in pairs, 7:34
  honey roasted, and airlines,
    4:13–14
  origins of comics name,
    10:191–193
Pear juice in canned peaches,
  8:46–47
Pears and apples, discoloration of,
  4:171

Pebbles, spitting by fish of, **9**:174–175

"Peeping Tom," **11**:162

Pencils
architectural and art, grades of, **7**:173
carpenter's, shape of, **7**:27; **9**:290–291
color, **3**:108
numbering, **3**:109

Penguins
frostbite on feet, **1**:217–218
knees, **5**:160

Penicillin and diet, **8**:95–96

Penmanship of doctors, bad, **5**:201; **6**:221–225; **7**:232–233; **8**:235

Pennies
lettering on, **7**:5
smooth edges of, **1**:40–41
vending machines and, **3**:54–56

Pennsylvania Dept. of Agriculture, registration, baked goods, **2**:121–122

Penny loafers, origins of, **8**:43–44

Pens
disappearance of, **4**:199; **5**:222–223; **6**:260–261; **7**:230–231; **8**:234
holes in barrel of cheap, **4**:111
ink leakages in, **4**:112–113

Pepper
and salt, as condiments, **5**:201; **6**:225–226; **8**:235–236
and sneezing, **8**:61
white, source of, **2**:135–136

Pepsi-Cola, trademark location of, **5**:115–116

Perfume
color of, **9**:19
wrists and, **6**:90

Periods in telegrams, **3**:77–78

Permanent press settings on irons, **3**:186–187

Permanents, pregnancy and, **3**:170–171

Perpetual care, cemeteries and, **2**:221–222

"Peter out," **11**:163–164

Phantom limb sensations, amputees and, **1**:73–75

Pharmacists and raised platforms, **8**:5–7

Phillips screwdriver, origins of, **2**:206–207

Philtrums, purpose of, **6**:43; **8**:266–267

Photo processing, one-hour, length of black-and-white film and, **4**:39

Photographs
poses in, **6**:218
red eyes in, **4**:68–69
stars in space and, **10**:213–215

Photography
color of cameras, **7**:246
hand position of men in old, **7**:24–26
hands on chins in, **7**:203–207
Polaroid prints, flapping of, **7**:175–176
smiling in old photographs, **7**:19–23

Physical exams, back tapping during, **3**:145–146

"Pi" as geometrical term, **5**:80–81

Piano keys, number of, **10**:7–9

"Pig in a poke," **11**:25

Pigeons
baby, elusiveness of, **1**:254; **10**:253–254
loss of toes, **7**:166
whistling sound in flight, **7**:58

Police car beacons, colors on,
7:135–137
Police dogs, urination and defeca-
tion of, 3:67–68
Policemen and mustaches, 6:219;
7:218–220; 8:246–247; 9:278
Ponds
effect of moons on, 5:138–139
fish returning to dried, 3:15–
16; 10:256
ice formations on, 5:82–83
versus lakes, 5:29–30; 7:241
versus lakes, level of, 9:85–86
Poodles, wild, 6:207–209
Pool balls, dots on, 10:237–240
"Pop goes the weasel," 11:14
Popcorn
"gourmet" versus regular, 1:176
popping in-house, in movie theaters, 1:45–50
versus other corns, 3:142–143
Popes
name change of, 10:17–18
white skullcap of, 10:80–81
white vestments of, 10:79–80
Popping noise of wood, in fires,
4:10
Pork and beans, pork in, 2:19
"Port," 11:65–66
"Porthole," 11:65–66
Postage and ripped stamps, 8:62
Postage stamps
leftover perforations of, 4:179
taste of, 2:182
Postal Service, U.S., undeliverable
mail and, 5:13–14
Pot pies, vent holes in, 6:28
Potato chips
bags, impossibility of opening
and closing, 9:117–118
curvy shape, 9:115–116
green tinges on, 5:136–137;
6:275

price of, versus tortilla chips,
5:137–138
Potato skins, restaurants and,
3:12–13
Potatoes, baked, and steak houses,
6:127–129
Potholes, causes of, 2:27
"Potter's field," 11:198
Power lines
humming of, 9:165–168;
10:259
orange balls on, 4:18–19
Prefaces in books, versus intro-
ductions and forewords,
1:72–73
Pregnancy, permanents and,
3:170–171
Pregnant women, food cravings of,
10:183–185
Preserves, contents of, 6:140–
141
Press conferences, microphones
in, 2:11–12
"Pretty kettle of fish," 11:178
"Pretty picnic," 11:178
Pretzels, shape of, 6:91–92
Priests, black vestments and,
10:77–79
Priority mail, first class versus,
3:166–167
Prisoners and license plate manufacturing, 8:137–139
Prohibition, liquor production of
distilleries during, 9:54–56
Pronunciation, dictionaries and,
10:169–179
"P's and Q's," 11:88–89
Pubic hair
curliness of, 5:177–178
purpose of, 2:146; 3:242–243;
6:275–276
Public buildings, temperatures in,
8:184

302   MASTER INDEX OF IMPONDERABILITY

Public radio, low frequency numbers of, **10**:181–183
Pudding, film on, **6**:57
Punts, measurement of, in football, **3**:124–125
Purple
    Christmas tree lights, **6**:185–186; **9**:293; **10**:280
    paganism, **9**:292–93
    royalty and, **6**:45–46
"Put up your dukes," **11**:137
Putting, veering of ball toward ocean when, **6**:107–108

"Q.T.," **11**:59
"Qantas," spelling of, **8**:134–135
Q-Tips, origins of name, **6**:210–211
"Quack [doctor]," **11**:45
"Quarterback," **11**:138
Quarterbacks and exclamation, "hut," **6**:210
Quarter-moons versus half moons, **7**:72–73
Quarts and gallons, American versus British, **4**:114–115
Queen-size sheets, size of, **3**:87–88

Rabbit tests, death of rabbits in, **7**:69–71
Rabbits and nose wiggling, **5**:173–174
Racewalking, judging of, **9**:20–23
Racquetballs, color of, **4**:8–9
Radar and police speed detection, **8**:88–91
Radiators and placement below windows, **9**:128–130
Radio
    beeps before network news, **1**:166–167

FM, odd frequency numbers of, **10**:59–60
    lack of song identification, **7**:51–57
    public, low frequency numbers, **10**:181–183
Radio Shack and lack of cash registers, **5**:165–166
Radios
    battery drainage, **10**:259–260
    lingering sound of recently unplugged, **4**:47
Railroad crossings and "EXEMPT" signs, **5**:118–119
Railroads, width of standard gauges of, **3**:157–159
Rain
    butterflies and, **4**:63–64
    fish biting in, **10**:131–138
    measurement container for, **4**:161–163
    smell of impending, **6**:170–171; **7**:241
Raincoats, dry-cleaning of, **2**:216–217
"Raining cats and dogs," **11**:26
"Raise hackles," **11**:6–7
Raisins
    cereal boxes and, **2**:123
    seeded grapes and, **2**:218–219
Ranchers' boots on fence posts, **2**:77–81; **3**:243–245; **4**:231; **5**:247–248; **6**:268; **7**:250; **8**:253; **9**:298–299; **10**:251–253
Razor blades, hotel slots for, **2**:113
Razors, men's versus women's, **6**:122–123
"Read the riot act," **11**:89
"Real McCoys," **11**:164–165

Rear admiral, origins of term, **5**:25

Rear-view mirrors, day/night switch on automobile, **4**:185–186

Receipts, cash register, color of, **4**:143

Records, vinyl, speeds of, **1**:58–61

Recreational vehicles and wheel covers, **7**:153

Recyclable plastics, numbers on, **6**:155–156

Recycling of newspaper ink, **7**:139–140

Red
color of beef, **8**:160–161
eyes in photographs, **4**:68–69
paint on coins, **7**:117

"Red cent," **11**:191

"Red herring," **11**:185–186

Red lights, versus green lights, on boats and airplanes, **4**:152–153

"Red tape," **11**:187

Red Wings, Detroit, octopus throwing and, **9**:183–186

"Red-letter day," **11**:186–187

Redshirting in college football, **7**:46–48

Refrigeration of opened food jars, **6**:171–172

Refrigerators
location of freezers in, **2**:230–231
smell of new, **8**:91–92
thermometers in, **10**:85–87

Relative humidity, during rain, **1**:225–226

Remote controls, men versus women and, **6**:217; **7**:193–196

Repair shops, backlogs and, **4**:45–47

Restaurants
coffee, versus home-brewed, **7**:181
vertical rulers near entrances of, **7**:95–96

Restrooms, group visits by females to, **6**:217; **7**:183–192; **8**:237–238; **9**:277–278

Revolving doors, appearance of, in big cities, **4**:171–173

Reynolds Wrap, texture of two sides of, **2**:102

Rhode Island, origins of name, **5**:21–22

Ribbons, blue, **3**:57–58

Rice cakes, structural integrity of, **10**:9–11

Rice Krispies
noises of, **3**:165; **5**:244–245
profession of Snap!, **8**:2–4

"Right wing," **11**:116

"Rigmarole," **11**:81–82

Rings, nose, bulls and, **10**:147–148

Rinks, ice, temperature of resurfacing water in, **10**:196–198

"Riot Act [1716]," **11**:89

Roaches
automobiles and, **7**:3–4; **8**:256–257; **9**:298
position of dead, **3**:133–134; **8**:256
reactions to light, **6**:20–21

Roads
blacktop, coloring of, **5**:22–23
fastening of lane reflectors on, **5**:98–99
versus bridges, in freezing characteristics, **5**:193

Robes, black, and judges, **6**:190–192

Rocking in zoo animals, **10**:279

Rodents and water sippers, **6**:187

Roller skating rinks, music in, **2**:107–108; **6**:274–275

Rolls
  coldness of airline, **3**:52–53
  versus buns, **8**:65–66
Roman chariots, flimsiness of,
  10:105–107
Roman numerals
  calculations with, **3**:105–106
  copyright notices in movie
    credits and, **1**:214–216
  on clocks, **5**:237–238
Roofs, gravel on, **6**:153–154
Roosevelt, Teddy, and San Juan
  Hill horses, **4**:49
Roosters, crowing and, **3**:3
Root beer
  carbonation in, **1**:87–88
  foam of, **8**:93–94
Rubble, Betty
  nonappearance in Flintstones
    vitamins, **6**:4–5; **9**:285–286
  vocation of, **7**:173–174
Ruins, layers of, **2**:138–140
Rulers, vertical, in restaurant en-
  trances, **7**:95–96
Run amok, **11**:68
Runny noses
  cold weather and, **10**:146–
    147
  kids versus adults, **9**:89–90
Rust, dental fillings and, **10**:41–
  42
RVs and wheel covers, **7**:153
"Rx," **11**:59

"Sacked [fired]," **11**:75
Sacks, paper
  jagged edges on, **6**:117–118
  names on, **2**:166–167
Safety caps, aspirin, 100-count
  bottles of, **4**:62
Safety pins, gold versus silver,
  9:87–88
Saffron, expense of, **1**:129–130

Sailors, bell bottom trousers and,
  2:84–85
"Salad days," **11**:178
Salads, restaurant, celery in,
  6:218; **7**:207–209
Saloon doors in Old West, **3**:198
Salt
  and pepper, as table condi-
    ments, **5**:201; **6**:225–226;
    8:235–236
  in oceans, **5**:149–150
  packaged, sugar as ingredient
    in, **4**:99
  round containers and, **10**:149–
    150
  storage bins on highway and,
    10:216–218
  versus sand, to treat icy roads,
    2:12–13
Salutes, military, origins of,
  3:147–149
San Francisco, sourdough bread
  in, **2**:180–181
Sand
  in pockets of new jeans, **7**:152
  storage bins on highway and,
    10:216–218
  versus salt, to treat icy roads,
    2:12–13
Sandbags, disposal of, **10**:193–
  195
Sardines, fresh, nonexistence in
  supermarkets, **1**:76–78
"Sawbuck [ten-dollar bill],"
  11:119–120
Sawdust on floor of bars, **10**:118–
  120
Scabs, itchiness of, **5**:125–126
Scars, hair growth and, **2**:186
"Schneider," **11**:136
School clocks, backward clicking
  of minute hands in, **1**:178–
  179

7:221–222; **8**:228–230;
9:271–272
tied to autos, at weddings,
1:235–238
uncomfortable, and women,
1:62–64
untied laces in stores, **8**:40
wing-tip, holes in, **8**:44
"Shoofly pie," **11**:179
Shopping, female proclivity toward, **7**:180; **8**:205–209
Shopping malls, doors at entrance
of, **6**:180–181
"Short shrift," **11**:15
Shoulder straps and seat belts in
airplanes, **8**:141–142
Shredded Wheat packages, Niagara Falls on, **5**:100–101
"Shrift," **11**:15
Shrimp, baby, peeling and clean-
ing of, **5**:127
"Shrive," **11**:15
"Siamese twins," **11**:151
"Sic," as dog command, **5**:51
Side vents in automobile windows,
6:13–14
"Sideburns," **11**:126–127
Sidewalks
cracks on, **3**:176–178
glitter on, **7**:160; **10**:61–62
Silica gel packs in electronics
boxes, **6**:201
Silos, round shape of, **3**:73–74;
5:245; **10**:260–261
Silver fillings, rusting of, **10**:41–
42
Silverstone, versus Teflon, **2**:3
Singers, American accents of for-
eign, **4**:125–126
Single-needle stitching in shirts,
6:51
Sinks, overflow mechanisms on,
2:214–215

Siskel, Gene, versus Roger Ebert,
billing of, **1**:137–139
Skating music, roller rinks and,
2:107–108; **6**:274–275
Skating, figure, and dizziness,
5:33–35
Ski poles, downhill, **3**:69
"Skidoo," **11**:100
Skunks, smell of, **10**:88–91
Skyscrapers, bricks in, **6**:102–
103
Skytyping, versus skywriting,
9:17–18
Skywriting
techniques of, **9**:12–16
versus skytyping, **9**:17–18
Sleep
babies and, **6**:56–57
drowsiness after meals, **6**:138–
139
eye position, **6**:146
heat and effect on, **6**:137–
138
twitching during, **2**:67
"Slippery When Wet" signs, location of, **7**:167–168
"Small fry," **11**:179
Smell of impending rain, **6**:170–
171; **7**:241
Smiling in old photographs, **7**:19–
23
Smoke from soda bottles, **8**:148
Snack foods and prepricing, **3**:79–
80
Snake emblems on ambulances,
6:144–145; **7**:239–240
Snakes
sneezing, **7**:98
tongues, **2**:106; **10**:278
Snap! [of Rice Krispies], profession of, **8**:2–4
"Snap! Crackle! And Pop!" of Rice
Krispies, **3**:165

Sneezing
    eye closure during, **3**:84–85
    looking up while, **2**:238
    pepper and, **8**:61
    snakes and, **7**:98
Snickers, wavy marks on bottom
    of, **6**:29–31
Sniffing, boxers and, **2**:22–23;
    10:256–258
Snoring, age differences and,
    10:126–127
Snow and cold weather, **3**:38
"Snow" on television, **9**:199–200
Soap, Ivory, purity of, **2**:46
Soaping of retail windows, **9**:265–
    267
Soaps, colored, white suds and,
    4:132–133
Social Security cards, lamination
    of, **5**:140–141
Social Security numbers
    fifth digit of, **8**:100–102
    reassignment of, **5**:61–62
    sequence of, **3**:91–92; **6**:267
Socks
    angle of, **3**:114–115
    disappearance of, **4**:127–128;
    5:245–246; **6**:272; **8**:266
    men's, coloring of toes on,
    4:19–20
"Soda jerk," **11**:176
Soft drinks
    bottles, holes on bottom of,
    6:187–188
    brominated vegetable oil in,
    6:96–97
    calorie constituents, **6**:94–95
    effect of container sizes and
    taste, **2**:157–159
    filling of bottles of, **4**:53
    finger as fizziness reduction
    agent, **9**:28–29
    fizziness in plastic cups, **9**:26

fizziness of soda with ice cream,
    9:27
fizziness over ice, **9**:24–25
freezing of, in machines, **9**:10–
    11
Kool-Aid and metal containers,
    8:51
machines, "Use Correct
    Change" light on, **9**:186–188
phenylalanine as ingredient in,
    6:96–97
pinholes on bottle caps of,
    · **2**:223
root beer, foam of, **8**:93–94
smoke of, **8**:148
"sodium-free" labels, **4**:87–88
Soles, sunburn and, **8**:63–64
"Son of a gun," **11**:82
Sonic booms, frequency of, **4**:23
Sororities, Greek names of,
    10:94–98
SOS pads, origins of, **8**:103–104
Soufflés and reaction to loud
    noises, **7**:41–42
Soup, alphabet, foreign countries
    and, **10**:73
Soups and shelving in supermar-
    kets, **6**:26–27
Sour cream, expiration date on,
    3:132
Sourdough bread, San Francisco,
    taste of, **2**:180–181
South Florida, University of, loca-
    tion of, **4**:7–8
South Pole
    directions at, **10**:243
    telling time at, **10**:241–243
Sparkling wine, name of, versus
    champagne, **1**:232–234
Speech, elderly versus younger
    and, **6**:24–25
Speed limit, 55 mph, reasons for,
    2:143

Speeding and radar, **8**:88–91

Speedometers, markings of, in automobiles, **2**:144–145

Spelling, "i" before "e" in, **6**:219; **7**:209; **8**:240–245

Sperm whales, head oil of, **6**:87–89

"Spic and span," **11**:41

Spiders and web tangling, **7**:169–170

Spitting, men versus women, **7**:181–182; **8**:226–227

Spoons, measuring, inaccuracy of, **1**:106–107

Sprinkles, jimmies versus, **10**:165–168

Squeaking, causes of, **9**:205–207

"St. Martin's Day," **11**:146

Stage hypnotists, techniques of, **1**:180–191

Staining, paperback books and, **2**:93–94

Stains, elimination of, on clothing, **6**:77–78

Staling of bread, **7**:125–126

Stamp pads, moisture retention of, **3**:24

Stamps
    perforation remnants, **4**:179
    postage, taste of, **2**:182
    validity of ripped, **8**:62

Staplers
    fitting of staples into, **10**:187–189
    outward setting of, **10**:189–190

Staples
    clumping of, **10**:46
    fitting into staplers of, **10**:187–189

"Starboard," **11**:65–66

Starch on shirts, **3**:118

Stars in space, photos of, **10**:213–215

Starving children and bloated stomachs, **7**:149–150

States, versus commonwealths, **7**:119–121

Static electricity, variability in amounts of, **4**:105–106

Steak houses and baked potatoes, **6**:127–129

Steak, "New York," origins of, **7**:155–156; **8**:252

Steam and streets of New York City, **5**:16–17

Steelers, Pittsburgh, helmet emblems of, **7**:67–68

Steins, beer, lids and, **9**:95–96

Stickers, colored, on envelopes, **4**:83–84

Stickiness of peanut butter, **10**:204–207

Stock prices as quoted in eighths of a dollar, **3**:112–113

Stocking runs
    direction of, **2**:124–125
    effect of freezing upon, **2**:125–126

"Stolen thunder," **11**:120

Stomach, growling, causes of, **4**:120–121

Stomachs, bloated, in starving children, **8**:254

"STOP" in telegrams, **3**:76–77

Strait, George, and hats, **6**:274

Straws, rising and sinking of, in glasses, **5**:36–37

Street addresses, half-numbers in, **8**:253

Street names, choice of, at corner intersections, **1**:154–156

Street-name signs, location of, at intersections, **1**:136

Thermometers
color of liquid in, **6**:147–148
in ovens and refrigerators,
10:85–87
placement of, **5**:158–159
standing temperature of, **1**:35–
37
Thimbles, holes in, **10**:63–64
Thinking, looking up and, **1**:55–58
"Third degree," **11**:93–94
"Third world," **11**:151–152
Thread, orange, and blue jeans,
9:74
Three Musketeers, musketlessness
. of, **7**:29–30
"Three sheets to the wind," **11**:94
Three-way light bulbs
burnout of, **2**:104
functioning of, **2**:105
Throat, uvula, purpose of, **3**:129
Throwing, sex differences in,
3:42–44; **6**:273–274
Thumb notches in dictionaries,
5:167–168
Thumbs-up gesture, "okay" and,
1:209–210
Thursday, Thanksgiving and,
4:140–142
Tickets, red carbons on airline,
3:179–180
Tickling of self, **4**:125
Ticklishness of different parts of
body, **10**:240–241
Ticks, diet of, **8**:68
Ties, origins of, **4**:127; **8**:264–265
Tiles, ceramic, in tunnels, **6**:135–
136; **8**:257–258
"Tinker's dam," **11**:82–83
Tinnitus, causes of, **2**:115–116
Tips, waiter, and credit cards,
. **8**:263
Tiredness and eye-rubbing,
10:103–105

Tires
atop mobile homes, **6**:163–164
automobile tread, disappearance of, **2**:72–74
bicycle, **2**:224–226
bluish tinge on new whitewalls,
6:192–193
inflation of, and gasoline
mileage, **6**:193–195
white-wall, size of, **2**:149
Tissue paper in wedding invitations, **4**:116–117
Title pages, dual, in books,
1:139–141
Toads, warts and, **10**:121–123
"Toady," **11**:83
"Toast [salute]," **11**:182
Toasters, one slice slot on, **3**:183–185
Toenails, growth of, **3**:123
Toffee, versus caramels, **1**:64
Toilet paper, folding over in hotel
bathrooms, **3**:4
Toilet seats in public restrooms,
3:137–138
Toilets
flush handles on, location of,
2:195–196
loud flushes of, in public rest-
rooms, **4**:187
seat covers for, **2**:83
"Tom Collins [drink]," **11**:165
"Tommy gun," **11**:166
Tongues, sticking out of, **3**:199;
7:224–225
Tonsils, function of, **5**:152–153
Toothpaste
expiration dates on, **4**:169–170;
6:266
taste of, with orange juice,
10:244–246
Toothpicks
flat versus round, **1**:224–225;
3:237–238

Valve stems on fire hydrants, shape of, **4**:142–143

VCRs, counter numbers on, **4**:148–149

Vegetable oil, vegetables in, **6**:266

Vegetable oils, vegetables in, **4**:186

Vending machines
bill counting and, **6**:271–272
freezing of soft drinks in, **9**:10–11
half dollars and, **3**:56–57
pennies and, **3**:54–56
placement of snacks in, **9**:162–164
"Use Correct Change" light on, **9**:186–188

Vent windows, side, in automobiles, **6**:13–14

Vestments
color of Catholic priests', **10**:77–79
color of popes', **10**:79–80

Videocassette boxes, configuration of, **9**:35–36

Videocassette recorders
counters on, **4**:149–49; **6**:272–273; **7**:243–244
power surges and, **7**:244–245

Videotape recorders
"play" and "record" switches on, **5**:23–24
storms and, **5**:180–181

Videotape versus audiotape, two sides of, **3**:136–137

Videotapes, rental, two-tone signals on, **5**:144–145

Violin bows, white dots on frogs of, **4**:164–165; **9**:291

Virgin acrylic, **7**:97–98

Virgin olive oil, **3**:174–175

Vision, 20–20, **3**:143

Vitamins, measurement of, in foods, **6**:148–150

Voices
causes of high and low, **2**:70
elderly versus younger, **6**:24–25
perception of, own versus others', **1**:95–96

Volkswagen Beetles, elimination of, **2**:192–194

Vomiting and horses, **6**:111–112

"Waffling," **11**:183

Wagon wheels in film, movement of, **2**:183

Waiters' tips and credit cards, **7**:133–135

Walking the plank, pirates and, **9**:37–42; **10**:273–274

Walking, race, judging of, **9**:20–23

Wall Street Journal, lack of photographs in, **4**:41–42

Warmth and its effect on pain, **3**:134–135

Warning labels, mattress tag, **2**:1–2

Warts, frogs and toads and, **10**:121–123

Washing machine agitators, movement of, **4**:56

Washing machines, top- versus bottom-loading, and detergent, **1**:159–165

Washington, D.C., "J" Street in, **2**:71

Watch, versus clock, distinctions between, **4**:77–78

"Watch," origins of term, **4**:77

Watches and placement on left hand, **4**:134–135; **6**:271

Water
bottled, expiration dates on, **9**:77–78
chemical manufacture of, **5**:107–108
clouds in tap water, **9**:126–127
color of, **2**:213

Water faucets
  bathroom versus kitchen,
    5:244
  location of, hot versus cold,
    4:191–192
Water temperature
  effect on stain, **6**:77–78
  versus air temperature, percep-
    tion of, **4**:184
Water towers
  height of, **5**:91–93
  winter and, **6**:38–40
Water, boiling during home births,
  6:114–115; **7**:247–248; **8**:254
Water, cold, kitchen versus bathroom, **4**:151–152
Watermelon seeds, white versus
  black, **5**:94
Wax, whereabouts in dripless can-
  dles, **4**:182–183
"Wear his heart on his sleeve,"
  11:128
"Weasel words," **11**:90–91
Weather
  clear days following storms,
    6:125
  forecasting of, in different re-
    gions, **6**:267–268
  partly cloudy versus partly
    sunny, **1**:21–22
  smell of impending rain,
    6:170–171
  West coast versus East coast,
    4:174–175
  See also particular conditions.
Wedding etiquette, congratulations
  to bride and groom and, **4**:86
Wedding invitations, tissue paper
  in, **4**:116–117
Weigh stations, highway, pre-
  dictable closure of, **4**:193–
  194
Wells, roundness of, **10**:228–229

Wendy's hamburgers, square
  shape of, **1**:113–115
Western Union telegrams
  exclamation marks and, **3**:76–77
  periods, **3**:77–78
Wet noses, dogs and, **4**:70–73
Wetness, effect of, on color, **4**:139
Whales, sperm, head oil in, **6**:87–89
Whiplash, delayed reaction of,
  10:128–130
Whips, cracking sound of, **2**:74
Whistling at sporting events, **3**:199
White Castle hamburgers, holes
  in, **1**:80–83
White chocolate, versus brown
  chocolate, **5**:134–135
"White elephant," **11**:191–192
White paint on homes, **2**:100–102
White pepper, source of, **2**:135–136
White wine, black grapes and,
  1:201–202
Whitewall tires
  bluish tinge on, **6**:192–193
  thickness of, **2**:149
Wigwams near highway, **10**:216–218
Wind on lakes, effect of different
  times on, **4**:156–157
Window cleaning, newspapers
  and, **10**:33–36
Window envelopes, use by mass
  mailers, **2**:111
Windows, rear, of automobiles,
  5:143–144
Windshield wipers, buses versus
  automobile, **7**:28
Wine
  chianti and straw-covered bot-
    tles, **8**:33–35
  dryness of, **5**:141–142
  temperature of serving, red ver-
    sus white, **4**:95–97
  white, black grapes and,
    1:201–202

Help!

We hate to end the book on a downbeat note, but we have to admit one dread fact: Imponderability is not yet smitten. Let's stamp it out.

"How?" you ask. Send us letters with your Imponderables, answers to Frustables, gushes of praise, and even your condemnations and corrections.

Join your inspired comrades and become a part of the wonderful world of *Imponderables*. If you are the first person to submit an Imponderable we use in the next volume, we'll send you a complimentary copy, along with an acknowledgment in the book.

Although we accept "snail mail," we strongly encourage you to e-mail us if possible. Because of the volume of mail, we can't always provide a personal response to every letter, but we'll try—a self-addressed stamped envelope doesn't hurt. We're much better with answering e-mail, although we fall far behind sometimes when work beckons.

Come visit us online at the Imponderables website, where you can pose Imponderables, read our blog, and find out what's happening at Imponderables Central. Send your correspondence, along with your name, address, and (optional) phone number to:

feldman@imponderables.com
www.imponderables.com

Imponderables
P.O. Box 116
Planetarium Station
New York, NY 10024-0116